**Lectures in Mathematics
ETH Zürich**
Department of Mathematics
Research Institute of Mathematics

Managing Editor:
Michael Struwe

Jean-François Le Gall

Spatial Branching Processes, Random Snakes and Partial Differential Equations

Springer Basel AG

Author's address:

Département de Mathématiques
Ecole Normale Supérieure
45, rue d'Ulm
F-75005 Paris

1991 Mathematical Subject Classification 60J80, 60J85, 60K35, 60H15

A CIP catalogue record for this book is available from the
Library of Congress, Washington D.C., USA

Deutsche Bibliothek Cataloging-in-Publication Data
Le Gall, Jean-François:
Spatial branching processes, random snakes and partial differential
equations / Jean-François Le Gall. - Basel ; Boston ; Berlin :
Birkhäuser, 1999
    (Lectures in mathematics : ETH Zürich)
    ISBN 978-3-7643-6126-6   ISBN 978-3-0348-8683-3 (eBooK)
    DOI 10.1007/978-3-0348-8683-3

© 1999 Springer Basel AG
Originally published by Birkhäuser Verlag in 1999

ISBN 978-3-7643-6126-6

9 8 7 6 5 4 3 2 1

Table of Contents

Foreword

In these lectures, we give an account of certain recent developments of the theory of spatial branching processes. These developments lead to several fascinating probabilistic objects, which combine spatial motion with a continuous branching phenomenon and are closely related to certain semilinear partial differential equations.

Our first objective is to give a short self-contained presentation of the measure-valued branching processes called superprocesses, which have been studied extensively in the last twelve years. We then want to specialize to the important class of superprocesses with quadratic branching mechanism and to explain how a concrete and powerful representation of these processes can be given in terms of the path-valued process called the Brownian snake. To understand this representation as well as to apply it, one needs to derive some remarkable properties of branching trees embedded in linear Brownian motion, which are of independent interest. A nice application of these developments is a simple construction of the random measure called ISE, which was proposed by Aldous as a tree-based model for random distribution of mass and seems to play an important role in asymptotics of certain models of statistical mechanics.

We use the Brownian snake approach to investigate connections between superprocesses and partial differential equations. These connections are remarkable in the sense that almost every important probabilistic question corresponds to a significant analytic problem. As Dynkin wrote in one of his first papers in this area, "it seems that both theories can gain from an interplay between probabilistic and analytic methods". A striking example of an application of analytic methods is the description of polar sets, which can be derived from the characterization of removable singularities of the corresponding partial differential equation. In the reverse direction, Wiener's test for the Brownian snake yields a characterization of those domains in which there exists a positive solution of $\Delta u = u^2$ with boundary blow-up. Both these results are presented in Chapter VI.

Although much of this book is devoted to the quadratic case, we explain in the last chapter how the Brownian snake representation can be extended to a general branching mechanism. This extension depends on certain deep connections with the theory of Lévy processes, which would have deserved a more thorough treatment.

Let us emphasize that this work does not give a comprehensive treatment of the theory of superprocesses. Just to name a few missing topics, we do not discuss the martingale problems for superprocesses, which are so important when dealing with regularity properties or constructing more complicated models, and we say nothing about catalytic superprocesses or interacting measure-valued processes. Even in the area of connections between superprocesses and partial differential equations, we leave aside such important tools as the special Markov property.

On the other hand, we have done our best to give a self-contained presentation and detailed proofs, assuming however some familiarity with Brownian motion and the basic facts of the theory of stochastic processes. Only in the last two chapters, we skip some technical parts of the arguments, but even there we hope that the important ideas will be accessible to the reader.

There is essentially no new result, even if we were able in a few cases to simplify the existing arguments. The bibliographical notes at the end of the book are intended to help the reader find his or her way through the literature. There is no claim for exhaustivity and we apologize in advance for any omission.

I would like to thank all those who attended the lectures, in particular Amine Asselah, Freddy Delbaen, Barbara Gentz, Uwe Schmock, Mario Wüthrich, Martin Zerner, and especially Alain Sznitman for several useful comments and for his kind hospitality at ETH. Mrs Boller did a nice job typing the first version of the manuscript. Finally, I am indebted to Eugene Dynkin for many fruitful discussions about the results presented here.

Paris, February 4, 1999

Frequently Used Notation

$\mathcal{B}(E)$ Borel σ-algebra on E.

$\mathcal{B}_b(E)$ Set of all real-valued bounded measurable functions on E.

$\mathcal{B}_+(E)$ Set of all nonnegative Borel measurable functions on E.

$\mathcal{B}_{b+}(E)$ Set of all nonnegative bounded Borel measurable functions on E.

$C(E)$ Set of all real-valued continuous functions on E.

$C_0(E)$ Set of all real-valued continuous functions with compact support on E.

$C_{b+}(E)$ Set of all nonnegative bounded continuous functions on E.

$C(E, F)$ Set of all continuous functions from E into F.

$C_0^\infty(\mathbb{R}^d)$ Set of all C^∞-functions with compact support on $\mathbb{R}^d$.

$\mathcal{M}_f(E)$ Set of all finite measures on E.

$\mathcal{M}_1(E)$ Set of all probability measures on E.

$\operatorname{supp}\mu$ Topological support of the measure μ.

$\dim A$ Hausdorff dimension of the set $A \subset \mathbb{R}^d$.

$B(x, r)$ Open ball of radius r centered at x.

$\bar{B}(x, r)$ Closed ball of radius r centered at x.

$\bar{A}$ Closure of the set A.

$\langle \mu, f \rangle = \int f \, d\mu$ for $f \in \mathcal{B}_+(E)$, $\mu \in \mathcal{M}_f(E)$.

$\operatorname{dist}(x, A) = \inf\{|y - x|, y \in A\},$ $x \in \mathbb{R}^d$, $A \subset \mathbb{R}^d$.

$p_t(x, y) = p_t(y - x) = (2\pi t)^{-d/2} \exp\left(-\dfrac{|y - x|^2}{2t}\right),$ $x, y \in \mathbb{R}^d$, $t > 0$.

$G(x, y) = G(y - x) = \int_0^\infty p_t(x, y) \, dt = \gamma_d |y - x|^{2-d},$ $x, y \in \mathbb{R}^d$, $d \geq 3$.

The letters C, c, c_1, c_2 etc. are often used to denote positive constants whose exact value is not specified.

Chapter I
An Overview

This first chapter gives an overview of the topics that will be treated in greater detail later, with pointers to the following chapters. We also discuss some recent related results which provide an a posteriori motivation for our investigations.

1 Galton-Watson processes and continuous-state branching processes

1.1 Galton-Watson processes are the simplest branching processes. They describe the evolution in discrete time of a population of individuals who reproduce themselves according to an offspring distribution μ. More precisely, starting from a probability measure μ on $\mathbb{N} = \{0, 1, 2, \ldots\}$, the associated Galton-Watson process is the Markov chain $(N_k, k \geq 0)$ with values in $\mathbb{N}$ such that, conditionally on N_n,

$$N_{n+1} \stackrel{(d)}{=} \sum_{i=1}^{N_n} \xi_i \, ,$$

where the variables ξ_i are i.i.d. with distribution μ and the symbol $\stackrel{(d)}{=}$ means equality in distribution.

Notice the obvious additivity (or branching) property: If $(N_k, k \geq 0)$ and $(N'_k, k \geq 0)$ are two independent Galton-Watson processes with offspring distribution μ, then so is $(N_k + N'_k, k \geq 0)$.

In what follows, we will concentrate on the critical or subcritical case, that is we assume

$$\sum_{k=0}^{\infty} k\mu(k) \leq 1 \, .$$

Then it is well known that the population becomes extinct in finite time: $N_k = 0$ for k large, a.s. (we exclude the trivial case when $\mu = \delta_1$ is the Dirac mass at 1).

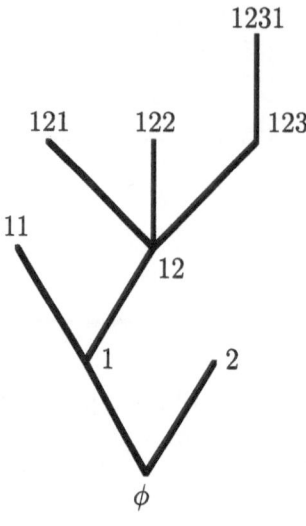

Fig. 1

The genealogy of a Galton-Watson process starting with 1 (resp. m) individuals at time 0 is obviously described by a random tree (resp. by m independent random trees) of the type of Fig. 1. Here we use the standard labelling of vertices (= individuals) of the tree. The ancestor is denoted by ϕ, the children of the ancestor by $1, 2, 3, \ldots$, the children of 1 by $11, 12, 13, \ldots$ and so on.

In view of forthcoming developments, it is important to realize that the knowledge of the tree provides more information than that of the associated Galton-Watson process (which corresponds to counting the number of individuals at every generation). For instance, a quantity such as the generation of the last common ancestor to individuals of the generation p is well-defined in terms of the tree but not in terms of the process $(N_k, k \geq 0)$.

1.2 Continuous-state branching processes are continuous analogues of the Galton-Watson branching processes. Roughly speaking, they describe the evolution in continuous time of a "population" with values in the positive real line $\mathbb{R}_+$. More precisely, we consider a Markov process $(Y_t, t \geq 0)$ with values in $\mathbb{R}_+$, whose sample paths are càdlàg, i.e. right-continuous with left limits. We say that Y is a continuous-state branching process (in short, a CSBP) if the transition kernels $P_t(x, dy)$ of Y satisfy the basic additivity property

$$P_t(x + x', \cdot) = P_t(x, \cdot) * P_t(x', \cdot) .$$

If we restrict ourselves to the critical or subcritical case (that is $\int P_t(x, dy) y \leq x$), one can then prove (Theorem II.1) that the Laplace transform of the transition kernel must be of the form

$$\int P_t(x, dy)e^{-\lambda y} = e^{-xu_t(\lambda)}$$

where the function $u_t(\lambda)$ is the unique nonnegative solution of the integral equation

$$u_t(\lambda) + \int_0^t \psi(u_s(\lambda))ds = \lambda$$

and ψ is a function of the following type

$$\psi(u) = \alpha u + \beta u^2 + \int_{(0,\infty)} \pi(dr)(e^{-ru} - 1 + ru) \tag{1}$$

where $\alpha \geq 0$, $\beta \geq 0$ and π is a σ-finite measure on $(0, \infty)$ such that $\int \pi(dr)(r \wedge r^2) < \infty$.

Conversely, for every function ψ of the previous type, there exists a (unique in law) continuous-state branching process Y associated with ψ (in short, a ψ-CSBP). The function ψ is called the branching mechanism of Y. In the formula for $\psi(u)$, the term αu corresponds to a killing at rate α (if $\psi(u) = \alpha u$, it is easy to see that $Y_t = Y_0 e^{-\alpha t}$), the measure π takes account of the jumps of Y (these jumps can only be positive), and the quadratic term βu^2 corresponds to a diffusion part.

In the special case when $\psi(u) = \beta u^2$ (quadratic branching mechanism), it is easy to compute

$$u_t(\lambda) = \frac{\lambda}{1 + \beta t \lambda}$$

and the process Y can be constructed as the solution of the stochastic differential equation

$$dY_t = \sqrt{2\beta Y_t}\, dB_t$$

where B is a one-dimensional Brownian motion (the well-known Yamada-Watanabe criterion shows that for every $y \geq 0$ the previous s.d.e. has a unique strong solution started at y, which is a continuous Markov process with values in $\mathbb{R}_+$). In this special case, Y is the so-called Feller diffusion, also known as the zero-dimensional squared Bessel process.

1.3 Continuous-state branching processes may also be obtained as weak limits of rescaled Galton-Watson processes. Suppose that, for every $k \geq 1$, we consider a Galton-Watson process $(N_n^k, n \geq 0)$ with initial value n_k and off-spring distribution μ_k possibly depending on k. If there exists a sequence of constants $a_k \uparrow \infty$ such that the rescaled processes

$$\left(\frac{1}{a_k} N_{[kt]}^k, \; t \geq 0\right)$$

converge to a limiting process $(Y_t, t \geq 0)$, at least in the sense of weak convergence of the finite-dimensional marginals, then the process Y must be a continuous-state branching process. Conversely, any continuous-state branching process can be obtained in this way (see Lamperti [La1] for both these results).

Of special interest is the case when $\mu_k = \mu$ for every k. Suppose first that μ is critical ($\sum k\mu(k) = 1$) with finite variance σ^2. Then the previous convergence holds with $a_k = k$ for every k, provided that $k^{-1}n_k \longrightarrow x$ for some $x \geq 0$. Furthermore, the limiting process is then a Feller diffusion, with $\psi(u) = \frac{1}{2}\sigma^2 u^2$. This result is known as the Feller approximation for branching processes.

More generally, when $\mu_k = \mu$ for every k, the limiting process Y (if it exists) must be of the stable branching type, meaning that

$$\psi(u) = cu^\gamma$$

for some $\gamma \in (1, 2]$. For $1 < \gamma < 2$, this corresponds to the choice $\alpha = \beta = 0$, $\pi(dr) = c'r^{-1-\gamma}dr$ in the previous formula for ψ.

1.4   We observed that the genealogy of a Galton-Watson process is described by a tree, or a finite collection of trees. A natural and important question is to get a similar description for the genealogy of a continuous-state branching process, which should involve some sort of continuous random tree. Furthermore, one expects that the genealogical trees of a sequence of Galton-Watson processes which converge after rescaling towards a continuous-state branching process should also converge in some sense towards the corresponding continuous genealogical structure. These questions will be discussed below.

2   Spatial branching processes and superprocesses

2.1   Spatial branching processes are obtained by combining the branching phenomenon with a spatial motion, which is usually given by a Markov process ξ with values in a Polish space E. In the discrete setting, the branching phenomenon is a Galton-Watson process, the individuals of generation n move between time n and time $n+1$ independently according to the law of ξ. At

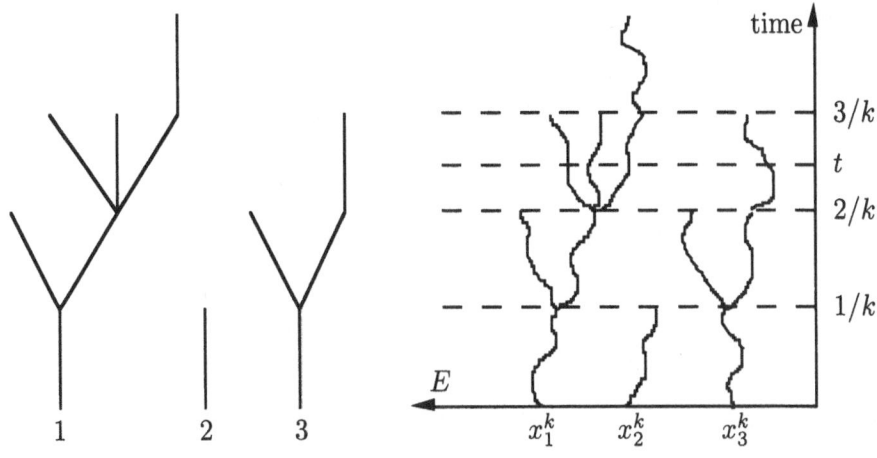

Fig. 2

time $n+1$ the newly born individuals start moving from the final position of their father, and so on (cf Fig. 2).

In the continuous setting, the branching phenomenon is a continuous-state branching process with branching mechanism ψ. The construction of the spatial motions is less easy but may be understood via the following approximation. As previously, consider a sequence N^k, $k \geq 1$ of Galton-Watson processes such that

$$\left(\frac{1}{a_k} N^k_{[kt]}, t \geq 0\right) \overset{(f.d.)}{\longrightarrow} (Y_t, t \geq 0) \tag{2}$$

where Y is a ψ-CSBP, and the symbol $\overset{(f.d.)}{\longrightarrow}$ means weak convergence of finite-dimensional marginals.

More precisely, we need for every k the genealogical trees associated with N^k. Then if $N^k_0 = n_k$, we consider n_k points $x^k_1, \ldots, x^k_{n_k}$ in E. We assume that the n_k initial individuals start respectively at $x^k_1, \ldots, x^k_{n_k}$ and then move independently according to the law of ξ between times $t = 0$ and $t = \frac{1}{k}$. At time $t = \frac{1}{k}$ each of these individuals is replaced by his children, who also move between times $t = \frac{1}{k}$ and $t = \frac{2}{k}$ according to the law of ξ, independently of each other, and so on.

Then, for every $t \geq 0$, let $\xi^{k,i}_t$, $i \in I(k,t)$ be the positions in E of the individuals alive at time t. Consider the random measure Z^k_t defined by

$$Z^k_t = \frac{1}{a_k} \sum_i \delta_{\xi^{k,i}_t} .$$

Then Z^k_t is a random element of the space $\mathcal{M}_f(E)$ of finite measures on E, which is equipped with the topology of weak convergence.

By construction the total mass of Z_t^k is

$$\langle Z_t^k, 1 \rangle = \frac{1}{a_k} N_{[kt]}^k \, ,$$

which by (2) converges to a ψ-CSBP.

Suppose that the initial values of Z^k converge as $k \to \infty$:

$$Z_0^k = \frac{1}{a_k} \sum_i \delta_{x_i^k} \longrightarrow \theta \in \mathcal{M}_f(E).$$

Then, under adequate regularity assumptions on the spatial motion ξ (satisfied for instance if ξ is Brownian motion in $\mathbb{R}^d$), there will exist an $\mathcal{M}_f(E)$-valued Markov process Z such that

$$(Z_t^k, t \geq 0) \longrightarrow (Z_t, t \geq 0)$$

in the sense of weak convergence of finite-dimensional marginals. The transition kernels of Z are characterized as follows. For $f \in \mathcal{B}_{b+}(E)$, and $s < t$,

$$E\big[\exp -\langle Z_t, f \rangle | Z_s \big] = \exp -\langle Z_s, v_{t-s} \rangle \, ,$$

where $(v_t(x), t \geq 0, x \in E)$ is the unique nonnegative solution of the integral equation

$$v_t(x) + \Pi_x \left(\int_0^t \psi(v_{t-s}(\xi_s)) \, ds \right) = \Pi_x \big(f(\xi_t) \big)$$

where we write Π_x for the probability measure under which ξ starts at x, and $\Pi_x U$ for the expectation of U under Π_x.

The process Z is called the (ξ, ψ)-superprocess. When ξ is Brownian motion in $\mathbb{R}^d$ and $\psi(u) = \beta u^2$, Z is called super-Brownian motion.

If ξ is a diffusion process in $\mathbb{R}^d$ with generator L, the integral equation for v_t is the integral form of the p.d.e.

$$\frac{\partial v_t}{\partial t} = L v_t - \psi(v_t) \, .$$

This provides a first connection between Z and p.d.e.'s associated with $Lu - \psi(u)$.

From our construction, or from the formula for the Laplace functional, it is clear that the total mass process $\langle Z_t, 1 \rangle$ is a ψ-CSBP.

2.2 What are the motivations for studying superprocesses?

In a sense, superprocesses are prototypes of infinite-dimensional Markov processes, for which many explicit calculations, concerning for instance hitting probabilities or moment functionals, are possible. The rich structure of superprocesses has allowed the derivation of many detailed sample path properties. A small sample of these will be given in Chapter IV.

There are interesting connections between superprocesses and stochastic partial differential equations. When ξ is linear Brownian motion and $\psi(u) = \beta u^2$, the measure Z_t has a density $z_t(x)$ (w.r.t. Lebesgue measure), which solves the equation

$$\frac{\partial z_t}{\partial t} = \frac{1}{2}\Delta z_t + \sqrt{2\beta z_t}\, \dot{W}_t$$

where W is space-time white noise (Konno and Shiga [KS], Reimers [R]). More general superprocesses with varying branching intensity can be used to construct solutions of more general s.p.d.e.'s (see Mueller and Perkins [MP], and the survey paper [DP2] for additional references).

The connections between superprocesses and partial differential equations associated with the operator $Lu - \psi(u)$ have been known for a long time and used in particular to understand the asymptotic behavior of superprocesses. In the beginning of the nineties, the introduction by Dynkin of exit measures, followed by the probabilistic solution of a nonlinear Dirichlet problem, led to a considerable progress in this area, which is still the object of active research. The connections between superprocesses and partial differential equations were initially used to get probabilistic information on superprocesses, but more recently they made it possible to prove new analytic results. This topic is treated in detail in Chapters V, VI, VII below.

One important initial motivation for superprocesses was the modelling of spatial populations. For this purpose it is often relevant to consider models with interactions. The construction of these more complicated models makes a heavy use of the technology developed to study superprocesses (see the survey paper [DP3] and the references therein). Catalytic superprocess, for which the branching phenomenon only occurs on a subset of the state space called the catalyst, can be thought of as modelling certain biological phenomena and have been studied extensively in the last few years (see the references in [DP3]).

Finally, remarkable connections have been obtained recently between superprocesses (especially super-Brownian motion) and models from statistical mechanics (lattice trees [DS], percolation clusters [HS]) or infinite particle systems (contact process [DuP], voter model [CDP], [BCL]). See the discussion in Section 6 below. This suggests that, like ordinary Brownian motion, super-Brownian motion is a universal object which arises in a variety of different contexts.

3   Quadratic branching and the Brownian snake

3.1   In the discrete setting we can construct the spatial branching process by first prescribing the branching structure (the genealogical trees) and then choosing the spatial motions (running independent copies of the process ξ along the branches of the tree). In order to follow the same route in the continuous setting, we need to describe the genealogical structure of a ψ-CSBP. We consider here the quadratic branching case $\psi(u) = \beta u^2$ and, to motivate the following construction, we recall a result due to Aldous.

We start from an offspring distribution μ which is critical and with finite variance, so that the corresponding Galton-Watson process suitably rescaled will converge to a CSBP with $\psi(u) = \beta u^2$. Under a mild assumption on μ, we can for every n sufficiently large define the law of the Galton-Watson tree with offspring distribution μ, conditioned to have a total population equal to n.

Aldous [Al3] proved that, provided that we rescale the lengths of the branches by the factor $\frac{c}{\sqrt{n}}$ (for a suitable choice of $c > 0$), these conditioned trees converge to the so-called continuum random tree (CRT).

To explain this result, we need

(A) to say what the CRT is;

(B) to say in which sense the convergence holds.

To begin with (A), we briefly explain how a continuous tree can be coded by a function. We consider a continuous function $e : [0, \sigma] \to \mathbb{R}_+$ such that $e(0) = e(\sigma) = 0$.

The key idea is that each $s \in [0, \sigma]$ corresponds to a vertex of the associated tree, but we identify s and t ($s \sim t$) if

$$e(s) = e(t) = \inf_{[s,t]} e(r)$$

(in particular $0 \sim \sigma$).

The quotient set $[0, \sigma]/\!\sim$ is the set of vertices. It is equipped with the partial order $s \prec t$ (s is an ancestor of t) iff

$$e(s) = \inf_{[s,t]} e(r)$$

and with the distance

$$d(s, t) = e(s) + e(t) - \inf_{[s,t]} e(r) \ .$$

Finally, the generation of the vertex s is $d(0, s) = e(s)$.

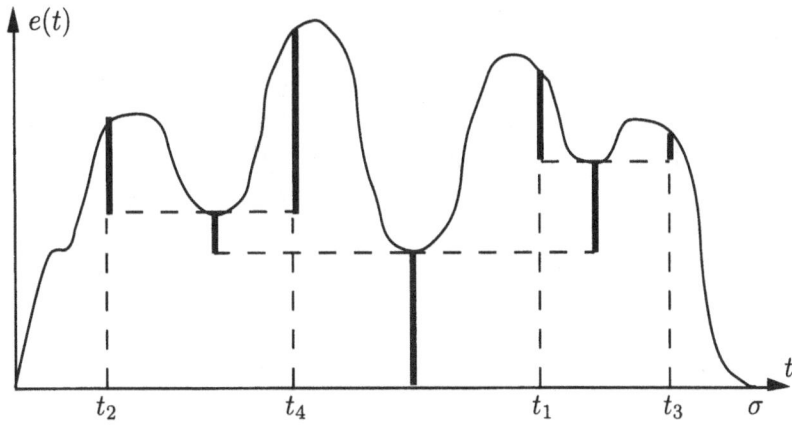

Fig. 3

The set of ancestors of a given vertex s is then isometric to the line segment $[0, e(s)]$. If $0 \leq s \leq t \leq \sigma$, the lines of ancestors of s and t have a common part isometric to the segment $[0, \inf_{[s,t]} e(r)]$ and then separate, etc. See Fig. 3 for the reduced tree consisting of the ancestors of four vertices $t_1, \ldots, t_4$ (note that, as far as the tree is concerned, horizontal distances have no meaning).

By definition, the CRT is the random tree obtained via the previous coding when e is chosen according to the law of the normalized Brownian excursion (i.e. the positive excursion of linear Brownian motion conditioned to have duration 1).

Concerning question (B), the convergence holds in the sense of convergence of the finite-dimensional marginals. For $p \geq 1$ fixed, the marginal of order p of a discrete tree is the law of the reduced tree consisting only of the ancestors of p individuals chosen uniformly (and independently) on the tree. Similarly, for the limiting tree CRT, the marginal of order p consists of the reduced tree associated with p instants $t_1, \ldots, t_p$ chosen uniformly and independently over $[0, \sigma]$ (cf Fig. 3 for an example with $p = 4$).

A more concrete way to express the convergence is to say that the (scaled) contour process of the discrete tree converges in distribution towards the normalized Brownian excursion. The contour process of the discrete tree is defined in the obvious way (cf Fig. 4) and it is scaled so that it takes a time $\frac{1}{2n}$ to visit any given edge.

A thorough discussion of the genealogical structure associated with Brownian excursions is presented in Chapter III.

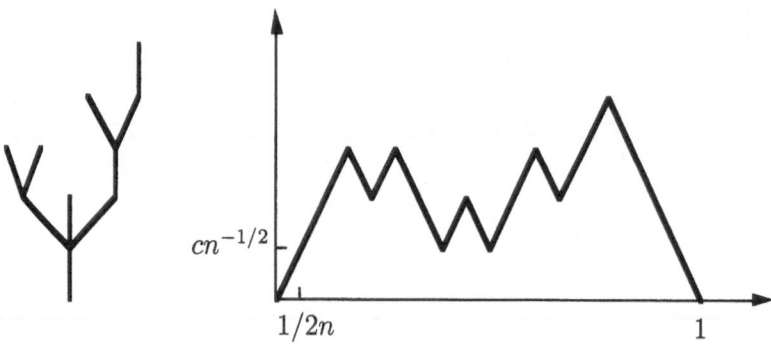

Fig. 4

3.2   The previous results strongly suggest that the genealogical structure of the Feller diffusion can be coded by Brownian excursions. This fact is illustrated by the Brownian snake construction of superprocesses with quadratic branching mechanism. The first step of this construction is to choose a collection of Brownian excursions, which will code the genealogical trees of the "population". In the second step, one constructs the spatial motions by attaching to each "individual" in these trees a path of the process ξ, in such a way that the paths of two individuals are the same up to the level corresponding to the generation of their last common ancestor.

To be specific, start from a reflected Brownian motion $(\zeta_s, s \geq 0)$ (ζ is distributed as the modulus of a standard linear Brownian motion started at 0). For every $a \geq 0$, denote by $(L_s^a, s \geq 0)$ the local time of ζ at level a. Then set $\eta_1 = \inf\{s \geq 0, L_s^0 > 1\}$. We will consider the values of ζ over the time interval $[0, \eta_1]$. By excursion theory, this means that we look at a Poisson collection of positive Brownian excursions.

Fix a point $y \in E$. Conditionally on $(\zeta_s, s \geq 0)$, we construct a collection $(W_s, s \geq 0)$ of finite paths in E so that the following holds.

(i) For every $s \geq 0$, W_s is a finite path in E started at y and defined on the time interval $[0, \zeta_s]$. (In particular W_0 is the trivial path consisting only of the starting point y.)

(ii) The mapping $s \to W_s$ is Markovian and if $s < s'$,

  - $W_{s'}(t) = W_s(t)$ if $t \leq \inf_{[s,s']} \zeta_r =: m(s, s')$;
  - conditionally on $W_s\big(m(s, s')\big)$, the path $\big(W_{s'}\big(m(s, s') + t\big), 0 \leq t \leq \zeta_{s'} - m(s, s')\big)$ is independent of W_s and distributed as the process ξ started at $W_s\big(m(s, s')\big)$, stopped at time $\zeta_{s'} - m(s, s')$.

Property (ii) is easy to understand if we think of s and s' as labelling two individuals in the tree who have the same ancestors up to generation $m(s, s')$:

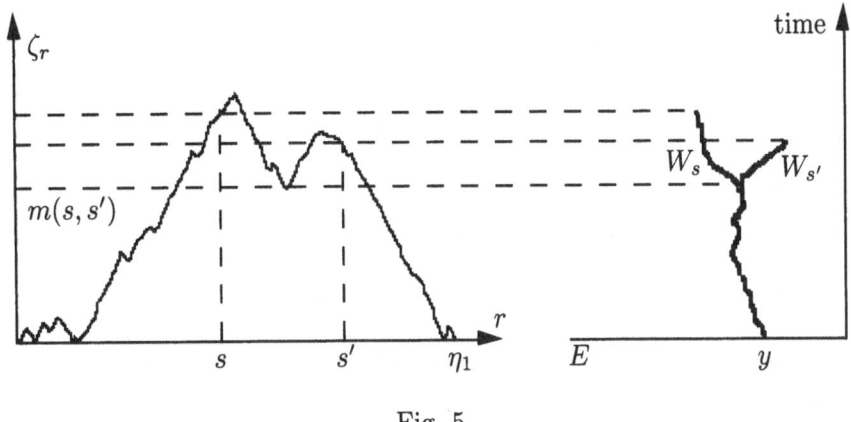

Fig. 5

Their spatial motions must be the same up to level $m(s, s')$, and then behave independently.

At an informal level one should view W_s as a path of the spatial motion ξ with a random lifetime ζ_s evolving like (reflecting) linear Brownian motion. The path gets "erased" when ζ_s decreases, and is extended when ζ_s increases.

The next theorem again requires some regularity of the spatial motion. A precise and more general statement is given in Chapter IV.

Theorem. *For every $a \geq 0$ let Z_a be the random measure on E defined by*

$$\langle Z_a, f \rangle = \int_0^{\eta_1} dL_s^a f(W_s(a))$$

where in the right side we integrate with respect to the increasing function $s \to L_s^a$. The process $(Z_a, a \geq 0)$ is a (ξ, ψ)-superprocess started at δ_y, with $\psi(u) = 2u^2$.

To interpret this theorem, we may say that the paths W_s, $0 \leq s \leq \eta_1$, are the "historical" paths of the "individuals" in a superprocess Z started at δ_y. For each $a \geq 0$, the support of Z_a is the set $\{W_s(a), 0 \leq s \leq \eta_1\}$ of positions of these paths at time a. The local time measure dL_s^a is used to construct Z_a as a measure "uniformly spread" over this set.

The process $(W_s, s \geq 0)$ is called the ξ-Brownian snake with initial point y. In what follows we will have to consider various choices of the initial point and we will use the notation $\mathbb{P}_y$ for the probability under which the initial point is y.

4   Some connections with partial differential equations

We will now discuss certain connections between superprocesses and a class of semilinear partial differential equations. We will assume that the spatial motion ξ is Brownian motion in $\mathbb{R}^d$ and we will rely on the Brownian snake construction.

Let D be a bounded domain in $\mathbb{R}^d$ and $y \in D$. If w is a finite path started at y, we set

$$\tau(w) = \inf\{t \geq 0, w(t) \notin D\} \leq \infty .$$

In a way analogous to the classical connections between Brownian motion and the Laplace equation $\Delta u = 0$, we are interested in the set of exit points

$$\mathcal{E}^D = \{W_s(\tau(W_s)) \; ; \; 0 \leq s \leq \eta_1 , \; \tau(W_s) < \infty\} .$$

Our first task is to construct a random measure that is in some sense uniformly distributed over $\mathcal{E}^D$.

Proposition. $\mathbb{P}_y$ *a.s. the formula*

$$\langle \mathcal{Z}^D, f \rangle = \lim_{\varepsilon \downarrow 0} \frac{1}{\varepsilon} \int_0^{\eta_1} f(W_s(\tau(W_s))) 1_{\{\tau(W_s) < \zeta_s < \tau(W_s) + \varepsilon\}} \, ds , \qquad f \in C(\partial D)$$

defines a random measure on ∂D called the exit measure from D.

The exit measure leads to the solution of the Dirichlet problem for the operator $\Delta u - u^2$ due to Dynkin (cf Chapter V for a proof, and Chapter VI for a number of applications). Note that we give in Chapter V a slightly different formulation in terms of excursion measures of the Brownian snake (a similar remark applies to the other results of this section).

Theorem. *Assume that ∂D is smooth and that f is continuous and nonnegative on ∂D. Then*

$$u(y) = -\log \mathbb{E}_y(\exp -\langle \mathcal{Z}^D, f \rangle) , \qquad y \in D$$

is the unique nonnegative solution of the problem

$$\begin{cases} \frac{1}{2}\Delta u = 2u^2 & \text{in } D , \\ u|_{\partial D} = f . \end{cases}$$

This theorem is the key to many other connections between super-Brownian motion or the Brownian snake and positive solutions of the p.d.e. $\Delta u = u^2$.

We will state here two of the corresponding results, which both lead to new analytic statements. The first theorem provides a classification of all positive solutions of $\Delta u = u^2$ in a smooth planar domain. A proof is provided in Chapter VII in the case when D is the unit disk.

Theorem. *Assume that $d = 2$ and ∂D is smooth. There is a one-to-one correspondence between*

- *nonnegative solutions of $\Delta u = 4u^2$ in D*
- *pairs (K, ν), where K is a compact subset of ∂D and ν is a Radon measure on $\partial D \backslash K$.*

If u is given,

$$K = \left\{ x \in \partial D \, , \, \limsup_{D \ni y \to x} \text{dist}(y, \partial D)^2 u(y) > 0 \right\}$$

and

$$\langle \nu, f \rangle = \lim_{r \downarrow 0} \int_{\partial D \backslash K} \sigma(dx) f(x) u(x + rN_x) \, , \qquad f \in C_0(\partial D \backslash K),$$

where $\sigma(dx)$ denotes the Lebesgue measure on ∂D, and N_x is the inward-pointing unit normal to ∂D at x. Conversely,

$$u(y) = - \log \mathbb{E}_y \left(1_{\{\mathcal{E}^D \cap K = \phi\}} \exp \left(- \int \nu(dx) z_D(x) \right) \right)$$

where $(z_D(x), x \in \partial D)$ is the continuous density of $\mathcal{Z}^D$ with respect to $\sigma(dx)$.

In higher dimensions $(d \geq 3)$ things become more complicated. One can still define the trace (K, ν) of a solution but there is no longer a one-to-one correspondence between a solution and its trace. Interesting results in this connection have been obtained recently by Dynkin and Kuznetsov (see the discussion at the end of Chapter VII). The previous theorem (except for the probabilistic representation) has been rederived by analytic methods, and extended to the equation $\Delta u = u^p$, provided that $d \leq d_0(p)$, by Marcus and Véron [MV2].

It has been known for a long time that if ∂D is smooth and $p > 1$ there exists a positive solution of $\Delta u = u^p$ in D that blows up everywhere at the boundary. One may ask whether this remains true for a general domain D. Our next result gives a complete answer when $p = 2$. If $r < r'$ we set

$$C(x, r, r') = \{ y \in \mathbb{R}^d, r \leq |y - x| \leq r' \} \, .$$

For every compact subset K of $\mathbb{R}^d$, we also define the capacity

$$c_{2,2}(K) = \inf\{\|\varphi\|_{2,2}^2 \; ; \; \varphi \in C_c^\infty(\mathbb{R}^d),$$

$$0 \leq \varphi \leq 1 \text{ and } \varphi = 1 \text{ on a neighborhood of } K\},$$

where $\|\varphi\|_{2,2}$ is the norm in the Sobolev space $W^{2,2}(\mathbb{R}^d)$. The following theorem is proved in Chapter VI.

Theorem. *Let D be a domain in $\mathbb{R}^d$. The following statements are equivalent.*

(i) *There exists a positive solution u of the problem*

$$\begin{cases} \Delta u = u^2 & \text{in } D, \\ u|_{\partial D} = +\infty. \end{cases}$$

(ii) *$d \leq 3$, or $d \geq 4$ and for every $x \in \partial D$*

$$\sum_{n=1}^\infty 2^{n(d-2)} c_{2,2}\big(D^c \cap \mathcal{C}(x, 2^{-n}, 2^{-n+1})\big) = \infty.$$

(iii) *If $T = \inf\{s \geq 0 \; ; \; W_s(t) \notin D \text{ for some } t > 0\}$, then $\mathbb{P}_x(T = 0) = 1$ for every $x \in \partial D$.*

From a probabilistic point of view, the previous theorem should be interpreted as a Wiener criterion for the Brownian snake: It gives a necessary and sufficient condition for the Brownian snake (or super-Brownian motion) started at a point $x \in \partial D$ to immediately exit D. Until today, there is no direct analytic proof of the equivalence between (i) and (ii) (some sufficient conditions ensuring (i) have been obtained by Marcus and Véron).

To conclude this section, let us emphasize that Chapters V, VI, VII are far from giving an exhaustive account of the known connections between superprocesses (or Brownian snakes) and partial differential equations: See in particular Dynkin and Kuznetsov [DK7], Etheridge [Et], Iscoe and Lee [IL], Lee [Le] and Sheu [Sh2], [Sh3] for interesting contributions to this area, which will not be discussed here.

5   More general branching mechanisms

The Brownian snake construction described in the previous sections relied on the coding of the genealogical structure of a CSBP with $\psi(u) = \beta u^2$ in terms of Brownian excursions. Our goal is now to extend this construction to more general branching mechanisms, of the type described above in Section 1.

To explain this extension, it is useful to start again from the discrete setting of Galton-Watson trees. Let μ be an offspring distribution and consider a sequence of independent Galton-Watson trees with offspring distribution μ. Then imagine a particle that "visits" successively all individuals of the different trees. For a given tree, individuals are visited in the lexicographical order. When the particle has visited all individuals of the first tree, it jumps to the ancestor of the second tree, and so on (cf. Fig. 6). For every $n \geq 0$ denote by the H_n the generation of the individual that is visited at time n.

 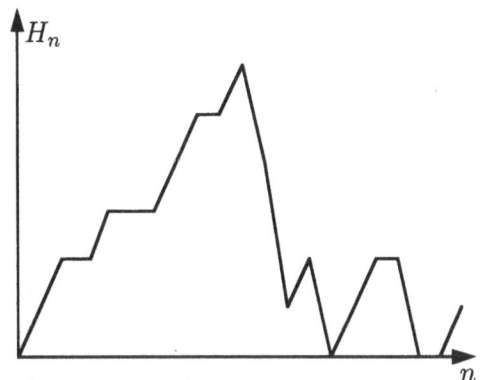

Fig. 6

The process $n \to H_n$ is called the height process corresponding to the offspring distribution μ. It is easy to see that the sequence of trees is completely determined by the function $n \to H_n$. In this sense we have described a coding of the sequence of trees (this coding is closely related, although different, to the contour process mentioned in Section 3).

At first glance, the previous coding does not seem particularly interesting, because the process H_n does not have nice properties (in particular it is usually not Markovian). The next proposition shows, however, that the height process is a simple functional of a random walk.

Proposition. *There exists a random walk on $\mathbb{Z}$, denoted by $(S_n, n \geq 0)$, with jump distribution $\nu(k) = \mu(k+1)$, $k = -1, 0, 1, 2, \ldots$, such that for every $n \geq 0$,*

$$H_n = \mathrm{Card}\{j \in \{0, 1, \ldots, n-1\}, S_j = \inf_{j \leq k \leq n} S_k\}.$$

(*Hint:* On the interval of visit of the m-th tree, the random walk S is defined by $S_n = U_n - (m-1)$, where U_n is the total number of "younger" brothers of the individual visited at time n and all his ancestors.)

Then let ψ be a branching mechanism function of the type (1), and let (μ_k) be a sequence of offspring distributions such that the corresponding Galton-Watson processes, suitably rescaled, converge in distribution towards the ψ-CSBP: As in Section 1, we assume that, if N^k is a Galton-Watson process with offspring distribution μ_k, started say at $N_0^k = a_k$, the convergence (2) holds and Y is a ψ-CSBP started at 1. For every k, let H^k be the height process corresponding to μ_k. We ask about the convergence of the (rescaled) processes H^k. The formula of the previous proposition suggests that the possible limit could be expressed in terms of the continuous analogue of the random walk S, that is a Lévy process with no negative jumps (observe that S has negative jumps only of size -1).

Let X be a Lévy process (real-valued process with stationary independent increments, started at 0) with Laplace exponent ψ:

$$E[\exp -\lambda X_t] = \exp(t\psi(\lambda)) , \quad \lambda \geq 0$$

(although X takes negative values, the Laplace transform $E[\exp -\lambda X_t]$ is finite for a Lévy process without negative jumps). Under our assumptions on ψ (cf (1)), X can be the most general Lévy process without negative jumps that does not drift to $+\infty$ as $t \to +\infty$.

We assume that the coefficient β of the quadratic part of ψ is strictly positive. For $0 \leq r \leq t$, we set

$$I_t^r = \inf_{r \leq s \leq t} X_s.$$

Theorem. ([LL1], [DuL]) *Under the previous assumptions, we have also*

$$\left(\frac{1}{k}H^k_{[ka_k t]}, t \geq 0\right) \xrightarrow[k \to \infty]{(f.d.)} (H_t, t \geq 0)$$

where the limiting process H is defined in terms of the Lévy process X by the formula

$$H_t = \beta^{-1} m(\{I_t^r ; 0 \leq r \leq t\})$$

if m denotes the Lebesgue measure on $\mathbb{R}$.

The process H_t is called the (continuous) height process. The formula for H_t is obviously a continuous analogue of the formula of the previous proposition. The theorem suggests that H codes the genealogy of the ψ-CSBP in the same way as reflected Brownian motion does when $\psi(u) = \beta u^2$. Indeed, we can observe that, when $\psi(u) = \beta u^2$, X is a (scaled) linear Brownian motion and

$$H_t = \frac{1}{\beta}\left(X_t - \inf_{o \leq s \leq t} X_s\right)$$

is a reflected Brownian motion, by a famous theorem of Lévy.

The Brownian snake construction of Section 3 extends to a general ψ, simply by replacing reflected Brownian motion by the process H. This extension is discussed in Chapter VIII. In the same spirit, one can use H to define a ψ-continuous random tree analogous to the CRT briefly described in Section 3. The finite-dimensional marginals of the ψ-continuous random tree can be computed explicitly in the stable case [DuL].

The convergence of finite-dimensional marginals in the previous theorem can be improved to a convergence in the functional sense, under suitable regularity assumptions. This makes it possible [DuL] to derive limit theorems for quantities depending on the *genealogy* of the Galton-Watson processes (for instance, the reduced tree consisting of ancestors of individuals of generation k). In this sense, the previous theorem shows that whenever a sequence of (rescaled) Galton-Watson processes converges, their genealogy also converges to the genealogy of the limiting CSBP.

6   Connections with statistical mechanics and interacting particle systems

In this last section, we briefly present two recent results which show that super-Brownian motion arises in a variety of different settings.

6.1   Lattice trees. A d-dimensional lattice tree with n bonds is a connected subgraph of $\mathbb{Z}^d$ with n bonds and $n+1$ vertices in which there are no loops. Let $Q_n(d\omega)$ be the uniform probability measure on the set of all lattice trees with n bonds that contain the origin. For every tree ω, let $X_n(\omega)$ be the probability measure on $\mathbb{R}^d$ obtained by putting mass $\frac{1}{n+1}$ to each vertex of the rescaled tree $cn^{-1/4}\omega$. Here $c = c(d)$ is a positive constant that must be fixed properly for the following to hold.

Following a conjecture of Aldous [Al4], Derbez and Slade [DS] proved that if d is large enough ($d > 8$ should be the right condition) the law of X_n under Q_n converges weakly as $n \to \infty$ to the law of the random measure $\mathcal{J}$ called ISE (Integrated Super-Brownian Excursion) which can be defined as follows.

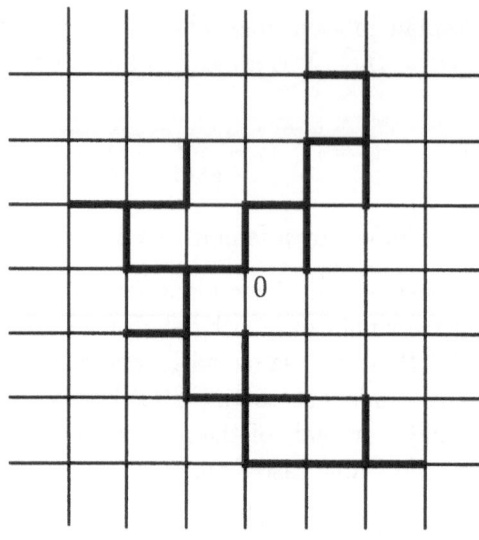

Fig. 7

Recall the Brownian snake construction of Section 3, in the special case when the spatial motion is d-dimensional Brownian motion and the initial point is 0. We use this construction, with the only difference that the lifetime process $(\zeta_s, 0 \leq s \leq 1)$ is a normalized Brownian excursion rather than reflected Brownian motion. If $(W_s, 0 \leq s \leq 1)$ is the resulting path-valued process, $\mathcal{J}$ may be defined by

$$\langle \mathcal{J}, f \rangle = \int_0^1 ds\, f(W_s(\zeta_s)) \,.$$

Alternatively, ISE can be viewed as combining a branching structure given by Aldous' CRT with Brownian motion in $\mathbb{R}^d$. A detailed discussion of ISE is presented in Chapter IV.

The proof of the Derbez-Slade result uses the lace expansion method developed by Brydges and Spencer. A work in progress of Hara and Slade [HS] also indicates that ISE arises as a scaling limit of the incipient infinite percolation cluster at the critical probability, again in high dimensions.

6.2   The voter model and coalescing random walks. The voter model is one of the most classical interacting particle systems. At each site $x \in \mathbb{Z}^d$ sits an individual who can have two possible opinions, say 0 or 1. At rate 1, each individual forgets his opinion and gets a new one by choosing one of his nearest neighbors uniformly at random, and taking his opinion. Our goal is to understand the way opinions propagate in space. For simplicity, we consider only the case $d \geq 3$.

Start from the simple situation where all individuals have type (opinion) 0 at the initial time, except for the individual at the origin who has type 1. Then with a high probability, type 1 will disappear. More precisely, if $\mathcal{U}_t$ denotes the set of individuals who have type 1 at time t, Bramson and Griffeath [BG] proved that

$$P[\mathcal{U}_t \neq \emptyset] \sim \frac{C}{t}$$

as $t \to \infty$. One may then ask about the shape of the set $\mathcal{U}_t$ conditional on the event $\{\mathcal{U}_t \neq \emptyset\}$. To state the result, let U_t be the random measure on $\mathbb{R}^d$ defined by

$$U_t = \frac{1}{t} \sum_{x \in \mathcal{U}_t} \delta_{x/\sqrt{t}} \, .$$

Then [BCL] the law of U_t conditionally on $\{\mathcal{U}_t \neq \emptyset\}$ converges as $t \to \infty$ to the law of $c\mathcal{H}$, where $c > 0$ and $\mathcal{H}$ is a random measure which is most conveniently described in terms of the Brownian snake as follows. Consider again the Brownian snake of Section 3 (with ξ Brownian motion in $\mathbb{R}^d$, initial point 0), but now assuming that the lifetime process $(\zeta_s, 0 \leq s \leq \sigma)$ is a Brownian excursion conditioned to hit level 1 ($\sup_{0 \leq s \leq \sigma} \zeta_s > 1$). Then

$$\langle \mathcal{H}, f \rangle = \int_0^\sigma dL_s^1 f\big(W_s(1)\big) \, ,$$

where L_s^1 is as previously the local time process of $(\zeta_s, 0 \leq s \leq \sigma)$ at level 1.

Alternatively, $\mathcal{H}$ can be described as super-Brownian motion at time 1 under its canonical measure. Closely related results showing that super-Brownian motion is the limit of rescaled voter models have been obtained by Cox, Durrett and Perkins [CDP].

A possible approach to the previous statement about the voter model is to use duality with a system of coalescing random walks. As a matter of fact, the result can be reformulated in terms of such a system. Suppose we start independent (simple) continuous-time random walks at every point x of $\mathbb{Z}^d$, and that any two random walks coalesce when they are at the same point at the same time. Let $\tilde{\mathcal{U}}_t$ be the set of all $x \in \mathbb{Z}^d$ such that the walk started at x is at 0 at time t (again $\tilde{\mathcal{U}}_t$ will be empty with a high probability) and let $\tilde{U}_t$ be the random measure

$$\tilde{U}_t = \frac{1}{t} \sum_{x \in \tilde{\mathcal{U}}_t} \delta_{x/\sqrt{t}} \, .$$

Then, the law of $\tilde{U}_t$ conditionally on $\{\tilde{\mathcal{U}}_t \neq \emptyset\}$ converges as $t \to \infty$ to the law of $c\mathcal{H}$.

A direct proof of the last result involves a careful analysis of the tree of coalescence for p coalescent random walks starting at different points of $\mathbb{Z}^d$. It turns out that the limiting behavior of this tree of coalescence can be described in terms of the genealogical structure of the Feller diffusion. This leads to the connection with super-Brownian motion or the Brownian snake.

Chapter II
Continuous-state Branching Processes and Superprocesses

In this chapter, we first obtain the general form of the Laplace functional of continuous-state branching processes, in the critical or subcritical case. We then provide a construction of these processes via an approximation by continuous-time Galton-Watson processes. If the branching phenomenon is combined with a spatial motion, a similar approximation leads to the measure valued processes called superprocesses. In the last two sections, we derive some basic properties of superprocesses.

1   Continuous-state branching processes

We consider a measurable family $\big(P_t(x,dy), t > 0, x \in \mathbb{R}_+\big)$ of transition kernels on the positive real line $\mathbb{R}_+$. This means that, for every $t > 0$ and $x \in \mathbb{R}_+$, $P_t(x,dy)$ is a probability measure on $\mathbb{R}_+$, the mapping $(t,x) \to P_t(x,A)$ is measurable for any Borel subset A of $\mathbb{R}_+$, and finally the Chapman-Kolmogorov equation $P_{t+s} = P_t P_s$ holds for every $t, s > 0$. We are interested in such families that satisfy the additivity or branching property $P_t(x,\cdot) * P_t(x',\cdot) = P_t(x + x',\cdot)$. The following theorem is a special case of a result due to Silverstein [Si].

Theorem 1. *Suppose that the family* $\big(P_t(x,dy), t > 0, x \in \mathbb{R}_+\big)$ *satisfies the following properties:*

(i) $P_t(x,\cdot) * P_t(x',\cdot) = P_t(x + x',\cdot)$ *for every* $t > 0$, $x, x' \in \mathbb{R}_+$.

(ii) $\int P_t(x,dy)\, y \le x$ *for every* $t > 0$, $x \in \mathbb{R}_+$.

Then, if we exclude the trivial case where $P_t(x,\cdot) = \delta_0$ *for every* $t > 0$ *and* $x \in \mathbb{R}_+$, *the Laplace functional of* $P_t(x,dy)$ *must be of the form*

$$\int P_t(x,dy)e^{-\lambda y} = e^{-x u_t(\lambda)}, \qquad \lambda \ge 0,$$

and the function $(u_t(\lambda), t \geq 0, \lambda \geq 0)$ *is the unique nonnegative solution of the integral equation*

$$u_t(\lambda) + \int_0^t ds\, \psi(u_s(\lambda)) = \lambda \,,$$

with a function ψ *of the form*

$$\psi(u) = \alpha u + \beta u^2 + \int \pi(dr)(e^{-ru} - 1 + ru) \,,$$

where $\alpha \geq 0$, $\beta \geq 0$ *and* π *is a* σ-*finite measure on* $(0, \infty)$ *such that* $\int \pi(dr)(r \wedge r^2) < \infty$.

Remark. A function ψ of the form given in the theorem is nonnegative and Lipschitz on compact subsets of $\mathbb{R}_+$. These properties play an important role in what follows.

Obviously if ψ is given, there is at most one associated family $(P_t(x, dy))$. We will see later that there is in fact exactly one.

Condition (ii) means that we consider only the critical or subcritical case. If we remove this condition, the theorem remains essentially true, with a more general form of ψ, but there are technical problems due to the possibility of explosion in finite time (for this reason, it is more convenient to consider transition kernels in $\bar{\mathbb{R}}_+ = \mathbb{R}_+ \cup \{+\infty\}$, see [Si]).

Proof. Assumption (i) implies that, for every fixed $t > 0$, $(P_t(x, \cdot), x \in \mathbb{R}_+)$ form a semigroup of infinitely divisible distributions on $\mathbb{R}_+$. By the Lévy-Khintchine formula, there exist $a_t \geq 0$ and a σ-finite measure n_t on $\mathbb{R}_+$, with $\int n_t(dr)(1 \wedge r) < \infty$, such that

$$\int P_t(x, dy)e^{-\lambda y} = e^{-x u_t(\lambda)} \,,$$

and

$$u_t(\lambda) = a_t \lambda + \int n_t(dr)(1 - e^{-\lambda r}) \,.$$

From (ii) and the Jensen inequality we have $u_t(\lambda) \leq \lambda$. By letting $\lambda \to 0$ we get

$$a_t + \int n_t(dr)r \leq 1 \,.$$

From the Chapman-Kolmogorov identity, we have also $u_{t+s} = u_t \circ u_s$. Since $u_t(\lambda) \leq \lambda$, we see that for every $\lambda \geq 0$ the mapping $t \to u_t(\lambda)$ is nonincreasing. By a standard differentiability theorem, the derivative

$$\lim_{s \to 0} \frac{u_{t+s}(\lambda) - u_t(\lambda)}{s} \tag{1}$$

exists for almost every $t > 0$, for every $\lambda \geq 0$.

We assumed that for some $t, x, P_t(x, \cdot) \neq \delta_0$. It follows that, for this value of t, $u_t(\lambda) > 0$. Using the relation $u_{t+s} = u_t \circ u_s$, we easily obtain that $u_t(\lambda) > 0$ for every $t > 0, \lambda > 0$.

By Fubini's theorem, we can pick $t_0 > 0$ such that the limit (1) exists for $t = t_0$ for a.a. $\lambda > 0$. Using the relation $u_{t_0+s}(\lambda) = u_s(u_{t_0}(\lambda))$ and the continuity of the mapping $\lambda \to u_{t_0}(\lambda)$ we get that

$$\lim_{s \downarrow 0} \frac{u_s(\gamma) - \gamma}{s} \tag{2}$$

exists for γ belonging to a dense subset of $(0, \gamma_0]$, where $\gamma_0 = u_{t_0}(1) > 0$. Fix any such γ, and observe that

$$\frac{\gamma - u_s(\gamma)}{s} = \left(1 - a_s - \int r\, n_s(dr)\right)\frac{\gamma}{s} + \frac{1}{s}\int (e^{-\gamma r} - 1 + \gamma r)n_s(dr) .$$

Notice that both terms in the right-hand side are nonnegative. The existence of the limit (2) implies that these terms are bounded when s varies over $(0, 1]$. It follows that there exists a constant C such that for every $s \in (0, 1]$,

$$\frac{1}{s}\int (r \wedge r^2)n_s(dr) \leq C , \quad \frac{1}{s}\left(1 - a_s - \int r\, n_s(dr)\right) \leq C .$$

By a standard compactness argument, there exists a sequence $s_k \downarrow 0$ such that the measures

$$\frac{1}{s_k}(r \wedge r^2)n_{s_k}(dr) + \frac{1}{s_k}\left(1 - a_{s_k} - \int r\, n_{s_k}(dr)\right)\delta_\infty(dr) \tag{3}$$

converge weakly to a finite measure $\eta(dr)$ on $[0, \infty]$.
Writing

$$\frac{\gamma - u_{s_k}(\gamma)}{s_k} = \left(1 - a_{s_k} - \int r\, n_{s_k}(dr)\right)\frac{\gamma}{s_k} + \frac{1}{s_k}\int \frac{e^{-\gamma r} - 1 + \gamma r}{r \wedge r^2}(r \wedge r^2)\, n_{s_k}(dr)$$

we get from weak convergence that for *every* $\gamma > 0$

$$\lim_{k \to \infty} \frac{\gamma - u_{s_k}(\gamma)}{s_k} = \alpha\gamma + \beta\gamma^2 + \int (e^{-\gamma r} - 1 + \gamma r)\pi(dr)$$

where $\alpha = \eta(\infty)$, $\beta = \frac{\eta(0)}{2}$ and $\pi(dr) = 1_{(0,\infty)}(r)(r \wedge r^2)^{-1}\eta(dr)$. Let $\psi(\gamma)$ denote the limit in the last displayed formula. Note that $\alpha = \psi'(0)$, and for $\gamma > 0$,

$$\psi(\gamma) = \alpha\gamma + \left(\beta + \int_0^\infty e^{-\gamma r}h(r)\, dr\right)\gamma^2,$$

where $h(r) = \int_r^\infty \pi([u, \infty))\, du$ is monotone decreasing and locally integrable over $\mathbb{R}_+$.

If we change the sequence (s_k), the limit $\psi(\gamma)$ must remain the same for γ belonging to a dense subset of $(0, \gamma_0]$, by (2). By standard arguments of analytic continuation, it follows that h, and then ψ and η, do not depend on the choice of the sequence (s_k). Therefore the convergence of the measures (3) holds as $s \downarrow 0$ and not only along the subsequence (s_k). We then conclude that, for every $\gamma \geq 0$,

$$\lim_{s \downarrow 0} \frac{u_s(\gamma) - \gamma}{s} = -\psi(\gamma)$$

and the limit is uniform when γ varies over compact subsets of $\mathbb{R}_+$. From the identity $u_{t+s}(\lambda) = u_s(u_t(\lambda))$ we see that the (right) derivative of $s \to u_s(\lambda)$ at t is $-\psi(u_t(\lambda))$, and the integral equation of the theorem now follows easily. $\square$

Exercise. With the notation of the previous proof, verify that $a_{t+s} = a_t a_s$. Then show that $a_t = 0$ for every $t > 0$ except possibly in the case when $\beta = 0$ and $\int r\pi(dr) < \infty$ [*Hint:* Assuming that $a_t > 0$ write

$$\frac{u_s(\gamma) - 1}{s} = \left(\frac{a_s - 1}{s}\right)\gamma + \frac{1}{s}\int (1 - e^{-\gamma r})n_s(dr)$$

and use the form of a_s to argue in a way similar to the proof of the theorem.]

We will now obtain the converse of Theorem 1. We fix a function ψ of the type introduced in Theorem 1, corresponding to the parameters α, β and ψ, and we will construct the associated family of transition kernels. To this end, we will use an approximation by continuous-time Galton-Watson processes. This approximation is useful to understand the behaviour of the Markov process associated with $P_t(x, dy)$, and especially the meaning of the parameters α, β and π.

We consider a Galton-Watson process in continuous time $X^\varepsilon = (X_t^\varepsilon, t \geq 0)$ where individuals die at rate $\rho_\varepsilon = \alpha_\varepsilon + \beta_\varepsilon + \gamma_\varepsilon$ (the parameters $\alpha_\varepsilon, \beta_\varepsilon, \gamma_\varepsilon \geq 0$ will be fixed later). When an individual dies, three possibilities may occur:

- with probability $\alpha_\varepsilon/\rho_\varepsilon$, the individual dies without descendants;
- with probability $\beta_\varepsilon/\rho_\varepsilon$, the individual gives rise to 0 or 2 children with probability $1/2$;
- with probability $\gamma_\varepsilon/\rho_\varepsilon$, the individual gives rise to a random number of offspring which is distributed as follows: Let V be a random variable distributed according to

$$\pi_\varepsilon(dv) = \pi\big((\varepsilon, \infty)\big)^{-1} 1_{\{v > \varepsilon\}} \pi(dv) ;$$

then, conditionally on V, the number of offspring is Poisson with parameter $m_\varepsilon V$, where $m_\varepsilon > 0$ is a parameter that will be fixed later.

In other words, the generating function of the branching distribution is:

$$\varphi_\varepsilon(r) = \frac{\alpha_\varepsilon}{\alpha_\varepsilon + \beta_\varepsilon + \gamma_\varepsilon} + \frac{\beta_\varepsilon}{\alpha_\varepsilon + \beta_\varepsilon + \gamma_\varepsilon}\left(\frac{1+r^2}{2}\right)$$

$$+ \frac{\gamma_\varepsilon}{\alpha_\varepsilon + \beta_\varepsilon + \gamma_\varepsilon}\int \pi_\varepsilon(dv)e^{-m_\varepsilon v(1-r)} .$$

We set $g_\varepsilon(r) = \rho_\varepsilon(\varphi_\varepsilon(r)-r) = \alpha_\varepsilon(1-r)+\frac{\beta_\varepsilon}{2}(1-r)^2+\gamma_\varepsilon(\int \pi_\varepsilon(dv)e^{-m_\varepsilon v(1-r)}-r)$. Write P_k for the probability measure under which X^ε starts at k. By standard results of the theory of branching processes (see e.g. Athreya-Ney [AN]), we have for $r \in [0,1]$,

$$E_1\left[r^{X_t^\varepsilon}\right] = v_t^\varepsilon(r)$$

where

$$v_t^\varepsilon(r) = r + \int_0^t g_\varepsilon\left(v_s^\varepsilon(r)\right)ds .$$

We are interested in scaling limits of the processes X^ε: We will start X^ε with $X_0^\varepsilon = [m_\varepsilon x]$, for some fixed $x > 0$, and study the behaviour of $m_\varepsilon^{-1}X_t^\varepsilon$. Thus we consider for $\lambda \geq 0$

$$E_{[m_\varepsilon x]}\left[e^{-\lambda m_\varepsilon^{-1}X_t^\varepsilon}\right] = v_t^\varepsilon\left(e^{-\lambda/m_\varepsilon}\right)^{[m_\varepsilon x]} = \exp\left([m_\varepsilon x]\log v_t^\varepsilon(e^{-\lambda/m_\varepsilon})\right) .$$

This suggests to define $u_t^\varepsilon(\lambda) = m_\varepsilon\left(1 - v_t^\varepsilon(e^{-\lambda/m_\varepsilon})\right)$. The function u_t^ε solves the equation

$$u_t^\varepsilon(\lambda) + \int_0^t \psi_\varepsilon\left(u_s^\varepsilon(\lambda)\right)ds = m_\varepsilon(1 - e^{-\lambda/m_\varepsilon}) , \tag{4}$$

where $\psi_\varepsilon(u) = m_\varepsilon g_\varepsilon(1 - m_\varepsilon^{-1}u)$.

From the previous formulas, we have

$$\psi_\varepsilon(u) = \alpha_\varepsilon u + m_\varepsilon^{-1}\beta_\varepsilon\frac{u^2}{2} + m_\varepsilon\gamma_\varepsilon\int \pi_\varepsilon(dr)(e^{-ru} - 1 + m_\varepsilon^{-1}u)$$

$$= \left(\alpha_\varepsilon - m_\varepsilon\gamma_\varepsilon\int \pi_\varepsilon(dr)r + \gamma_\varepsilon\right)u + m_\varepsilon^{-1}\beta_\varepsilon\frac{u^2}{2}$$

$$+ m_\varepsilon\gamma_\varepsilon\pi\left((\varepsilon,\infty)\right)^{-1}\int_{(\varepsilon,\infty)} \pi(dr)(e^{-ru} - 1 + ru) .$$

At the present stage we want to choose the parameters $\alpha_\varepsilon, \beta_\varepsilon, \gamma_\varepsilon$ and m_ε so that

(i) $\lim_{\varepsilon \downarrow 0} m_\varepsilon = +\infty$.

(ii) If $\pi \neq 0$, $\lim_{\varepsilon \downarrow 0} m_\varepsilon \gamma_\varepsilon \pi\big((\varepsilon, \infty)\big)^{-1} = 1$. If $\pi = 0$, $\gamma_\varepsilon = 0$.

(iii) $\lim_{\varepsilon \downarrow 0} \frac{1}{2} m_\varepsilon^{-1} \beta_\varepsilon = \beta$.

(iv) $\lim_{\varepsilon \downarrow 0}\big(\alpha_\varepsilon - m_\varepsilon \gamma_\varepsilon \int \pi_\varepsilon(dr) r + \gamma_\varepsilon\big) = \alpha$, and $\alpha_\varepsilon - m_\varepsilon \gamma_\varepsilon \int \pi_\varepsilon(dr) r + \gamma_\varepsilon \geq 0$, for every $\varepsilon > 0$.

Obviously it is possible, in many different ways, to choose $\alpha_\varepsilon, \beta_\varepsilon, \gamma_\varepsilon$ and m_ε such that these properties hold.

Proposition 2. *Suppose that properties (i)–(iv) hold. Then, for every $t > 0$, $x \geq 0$, the law of $m_\varepsilon^{-1} X_t^\varepsilon$ under $P_{[m_\varepsilon x]}$ converges as $\varepsilon \to 0$ to a probability measure $P_t(x, dy)$. Furthermore, the kernels $\big(P_t(x, dy), t > 0, x \in \mathbb{R}_+\big)$ are associated with the function ψ in the way described in Theorem 1.*

Proof. From (i)–(iv) we have

$$\lim_{\varepsilon \downarrow 0} \psi_\varepsilon(u) = \psi(u) \tag{5}$$

uniformly over compact subsets of $\mathbb{R}_+$. Let $u_t(\lambda)$ be the unique nonnegative solution of

$$u_t(\lambda) + \int_0^t \psi(u_s(\lambda)) ds = \lambda \tag{6}$$

($u_t(\lambda)$ may be defined by: $\int_{u_t(\lambda)}^\lambda \psi(v)^{-1} dv = t$ when $\lambda > 0$; this definition makes sense because $\psi(v) \leq Cv$ for $v \leq 1$, so that $\int_{0+} \psi(v)^{-1} dv = +\infty$).

We then make the difference between (4) and (6), and use (5) and the fact that ψ is Lipschitz over $[0, \lambda]$ to obtain

$$|u_t(\lambda) - u_t^\varepsilon(\lambda)| \leq C_\lambda \int_0^t |u_s(\lambda) - u_s^\varepsilon(\lambda)| \, ds + a(\varepsilon, \lambda)$$

where $a(\varepsilon, \lambda) \to 0$ as $\varepsilon \to 0$, and the constant C_λ is the Lipschitz constant for ψ on $[0, \lambda]$. We conclude from Gronwall's lemma that for every $\lambda \geq 0$,

$$\lim_{\varepsilon \downarrow 0} u_t^\varepsilon(\lambda) = u_t(\lambda)$$

uniformly on compact sets in t.

Coming back to a previous formula we have

$$\lim_{\varepsilon \to 0} E_{[m_\varepsilon x]}\left[e^{-\lambda m_\varepsilon^{-1} X_t^\varepsilon}\right] = e^{-x u_t(\lambda)}$$

and the first assertion of the proposition follows from a classical statement about Laplace transforms.

The end of the proof is straightforward. The Chapman-Kolmogorov relation for $P_t(x, dy)$ follows from the identity $u_{t+s} = u_t \circ u_s$, which is easy from (6). The additivity property is immediate since

$$\int P_t(x+x', dy)e^{-\lambda y} = e^{-(x+x')u_t(\lambda)} = \left(\int P_t(x, dy)e^{-\lambda y}\right)\left(\int P_t(x', dy)e^{-\lambda y}\right).$$

The property $\int P_t(x, dy)y \le x$ follows from the fact that

$$\limsup_{\lambda \to 0} \lambda^{-1} u_t(\lambda) \le 1.$$

Finally the kernels $P_t(x, dy)$ are associated with ψ by construction.   □

Definition. *The ψ-continuous-state branching process (in short, the ψ-CSBP) is the Markov process in $\mathbb{R}_+$ $(X_t, t \ge 0)$ whose transition kernels $P_t(x, dy)$ are associated with the function ψ by the correspondence of Theorem 1. The function ψ is called the branching mechanism of X.*

Examples.

  (i) If $\psi(u) = \alpha u$, $X_t = X_0 e^{-\alpha t}$

  (ii) If $\psi(u) = \beta u^2$ one can compute explicitely $u_t(\lambda) = \frac{\lambda}{1 + \beta \lambda t}$. The corresponding process X is called the Feller diffusion, for reasons that are explained in the exercise below.

  (iii) By taking $\alpha = \beta = 0$, $\pi(dr) = c\frac{dr}{r^{1+b}}$ with $1 < b < 2$, one gets $\psi(u) = c'u^b$. This is called the stable branching mechanism.

From the form of the Laplace functionals, it is very easy to see that the kernels $P_t(x, dy)$ satisfy the Feller property, as defined in [RY] Chapter III (use the fact that linear combinations of functions $e^{-\lambda x}$ are dense in the space of continuous functions on $\mathbb{R}_+$ that tend to 0 at infinity). By standard results, every ψ-CSBP has a modification whose paths are right-continuous with left limits, and which is also strong Markov.

Exercise. Verify that the Feller diffusion can also be obtained as the solution to the stochastic differential equation

$$dX_t = \sqrt{2\beta X_t} dB_t$$

where B is a one-dimensional Brownian motion [*Hint*: Apply Itô's formula to see that

$$\exp\left(-\frac{\lambda X_s}{1 + \beta\lambda(t - s)}\right), \quad 0 \leq s \leq t$$

is a martingale.]

Exercise. (Almost sure extinction) Let X be a ψ-CSBP started at $x > 0$, and let $T = \inf\{t \geq 0, X_t = 0\}$. Verify that $X_t = 0$ for every $t \geq T$, a.s. (use the strong Markov property). Prove that $T < \infty$ a.s. if and only if

$$\int^{\infty} \frac{du}{\psi(u)} < \infty.$$

(This is true in particular for the Feller diffusion.) If this condition fails, then $T = \infty$ a.s.

2   Superprocesses

In this section we will combine the continuous-state branching processes of the previous section with spatial motion, in order to get the so-called super-processes. The spatial motion will be given by a Markov process $(\xi_s, s \geq 0)$ with values in a Polish space E. We assume that the paths of ξ are càdlàg (right-continuous with left limits) and so ξ may be defined on the canonical Skorokhod space $\mathbb{D}(\mathbb{R}_+, E)$. We write Π_x for the law of ξ started at x. The mapping $x \rightarrow \Pi_x$ is measurable by assumption. We denote by $\mathcal{B}_{b+}(E)$ (resp. $C_{b+}(E)$) the set of all bounded nonnegative measurable (resp. bounded non-negative continuous) functions on E.

In the spirit of the previous section, we use an approximation by branching particle systems. Recall the notation ρ_ε, m_ε, φ_ε of the previous section. We suppose that, at time $t = 0$, we have N_ε particles located respectively at points $x_1^\varepsilon, \ldots, x_{N_\varepsilon}^\varepsilon$ in E. These particles move independently in E according to the law of the spatial motion ξ. Each particle dies at rate ρ_ε and gives rise to a random number of offspring according to the distribution with generating function φ_ε. Let Z_t^ε be the random measure on E defined as the sum of the Dirac masses at the positions of the particles alive at t. Our goal is to investigate the limiting behavior of $m_\varepsilon^{-1} Z_t^\varepsilon$, for a suitable choice of the initial distribution.

The process $(Z_t^\varepsilon, t \geq 0)$ is a Markov process with values in the set $M_p(E)$ of all point measures on E. We write $\mathbb{P}_\theta^\varepsilon$ for the probability measure under which Z^ε starts at θ.

Fix a Borel function f on E such that $c \leq f \leq 1$ for some $c > 0$. For every $x \in E$, $t \geq 0$ set

$$w_t^\varepsilon(x) = \mathbb{E}_{\delta_x}^\varepsilon \left(\exp\langle Z_t^\varepsilon, \log f \rangle \right)$$

where we use the notation $\langle \mu, f \rangle = \int f \, d\mu$.
Note that the quantity $\exp\langle Z_t^\varepsilon, \log f \rangle$ is the product of the values of f evaluated at the particles alive at t.

Proposition 3. *The function $w_t^\varepsilon(x)$ solves the integral equation*

$$w_t^\varepsilon(x) - \rho_\varepsilon \Pi_x \left(\int_0^t ds \big(\varphi_\varepsilon(w_{t-s}^\varepsilon(\xi_s)) - w_{t-s}^\varepsilon(\xi_s) \big) \right) = \Pi_x \left(f(\xi_t) \right).$$

Proof. Since the parameter ε is fixed for the moment, we omit it, only in this proof. Note that we have for every positive integer n

$$\mathbb{E}_{n \delta_x} \left(\exp\langle Z_t, \log f \rangle \right) = w_t(x)^n.$$

Under $\mathbb{P}_{\delta_x}$ the system starts with one particle located at x. Denote by T the first branching time and by M the number of offspring of the initial particle. Let also $P_M(dm)$ be the law of M (the generating function of P_M is φ). Then

$$w_t(x) = \mathbb{E}_{\delta_x} \left(1_{\{T > t\}} \exp\langle Z_t, \log f \rangle \right) + \mathbb{E}_{\delta_x} \left(1_{\{T \leq t\}} \exp\langle Z_t, \log f \rangle \right)$$

$$= e^{-\rho t} \Pi_x \left(f(\xi_t) \right) + \rho \, \Pi_x \otimes P_M \left(\int_0^t ds \, e^{-\rho s} \mathbb{E}_{m \delta_{\xi_s}} \left(\exp\langle Z_{t-s}, \log f \rangle \right) \right)$$

$$= e^{-\rho t} \Pi_x \left(f(\xi_t) \right) + \rho \, \Pi_x \left(\int_0^t ds \, e^{-\rho s} \varphi(w_{t-s}(\xi_s)) \right). \tag{7}$$

The integral equation of the proposition is easily derived from this identity: From (7) we have

$$\rho \, \Pi_x \left(\int_0^t ds \, w_{t-s}(\xi_s) \right)$$

$$= \rho \, \Pi_x \left(\int_0^t ds \, e^{-\rho(t-s)} \Pi_{\xi_s} \left(f(\xi_{t-s}) \right) \right)$$

$$+ \rho^2 \Pi_x \left(\int_0^t ds \, \Pi_{\xi_s} \left(\int_0^{t-s} dr \, e^{-\rho r} \varphi(w_{t-s-r}(\xi_r)) \right) \right)$$

$$= \rho \int_0^t ds \, e^{-\rho(t-s)} \Pi_x \left(f(\xi_t) \right)$$

$$+ \rho^2 \int_0^t ds \int_s^t dr \, e^{-\rho(r-s)} \Pi_x \left(\Pi_{\xi_s} \left(\varphi(w_{t-r}(\xi_{r-s})) \right) \right)$$

$$= (1 - e^{-\rho t}) \Pi_x \left(f(\xi_t) \right) + \rho \, \Pi_x \left(\int_0^t dr (1 - e^{-\rho r}) \varphi(w_{t-r}(\xi_r)) \right).$$

By adding this equality to (7) we get Proposition 3. □

We now fix a function $g \in \mathcal{B}_{b+}(E)$, then take $f = e^{-m_\varepsilon^{-1}}g$ in the definition of $w_t^\varepsilon(x)$ and set

$$u_t^\varepsilon(x) = m_\varepsilon\left(1 - w_t^\varepsilon(x)\right) = m_\varepsilon\left(1 - \mathbb{E}_{\delta_x}^\varepsilon\left(e^{-m_\varepsilon^{-1}\langle Z_t^\varepsilon, g\rangle}\right)\right).$$

From Proposition 3, it readily follows that

$$u_t^\varepsilon(x) + \Pi_x\left(\int_0^t ds\,\psi_\varepsilon\left(u_{t-s}^\varepsilon(\xi_s)\right)\right) = m_\varepsilon\Pi_x\left(1 - e^{-m_\varepsilon^{-1}g(\xi_t)}\right) \qquad (8)$$

where the function ψ_ε is as in Section 1.

Lemma 4. *Suppose that conditions (i)–(iv) before Proposition 2 hold. Then, the limit*

$$\lim_{\varepsilon \to 0} u_t^\varepsilon(x) =: u_t(x)$$

exists for every $t \geq 0$ and $x \in E$, and the convergence is uniform on the sets $[0, T] \times E$. Furthermore, $u_t(x)$ is the unique nonnegative solution of the integral equation

$$u_t(x) + \Pi_x\left(\int_0^t ds\,\psi\left(u_{t-s}(\xi_s)\right)\right) = \Pi_x\left(g(\xi_t)\right). \qquad (9)$$

Proof. From our assumptions, $\psi_\varepsilon \geq 0$, and so it follows from (8) that $u_t^\varepsilon(x) \leq \lambda := \sup_{x \in E} g(x)$. Also note that

$$\lim_{\varepsilon \to 0} m_\varepsilon\Pi_x\left(1 - e^{-m_\varepsilon^{-1}g(\xi_t)}\right) = \Pi_x\left(g(\xi_t)\right)$$

uniformly in $(t, x) \in \mathbb{R}_+ \times E$ (indeed the rate of convergence only depends on λ). Using the uniform convergence of ψ_ε towards ψ on $[0, \lambda]$ (and the Lipschitz property of ψ as in the proof of Proposition 2), we get for $\varepsilon > \varepsilon' > 0$ and $t \in [0, T]$,

$$|u_t^\varepsilon(x) - u_t^{\varepsilon'}(x)| \leq C_\lambda \int_0^t ds \sup_{y \in E} |u_s^\varepsilon(y) - u_s^{\varepsilon'}(y)| + b(\varepsilon, T, \lambda)$$

where $b(\varepsilon, T, \lambda) \to 0$ as $\varepsilon \to 0$. From Gronwall's lemma, we obtain that $u_t^\varepsilon(x)$ converges uniformly on the sets $[0, T] \times E$. Passing to the limit in (8) shows that the limit satisfies (9). Finally the uniqueness of the nonnegative solution of (9) is also a consequence of Gronwall's lemma. $\square$

We are now ready to state our basic construction theorem for superprocesses. We denote by $\mathcal{M}_f(E)$ the space of all finite measures on E, which is equipped with the topology of weak convergence.

Theorem 5. *For every $\mu \in \mathcal{M}_f(E)$ and every $t > 0$, there exists a (unique) probability measure $Q_t(\mu, d\nu)$ on $\mathcal{M}_f(E)$ such that for every $g \in \mathcal{B}_{b+}(E)$,*

$$\int Q_t(\mu, d\nu)e^{-\langle \nu, g \rangle} = e^{-\langle \mu, u_t \rangle} \tag{10}$$

where $(u_t(x), x \in E)$ is the unique nonnegative solution of (9). The collection $Q_t(\mu, d\nu)$, $t > 0$, $\mu \in \mathcal{M}_f(E)$ is a measurable family of transition kernels on $\mathcal{M}_f(E)$, which satisfies the additivity property

$$Q_t(\mu, \cdot) * Q_t(\mu', \cdot) = Q_t(\mu + \mu', \cdot) .$$

The Markov process Z in $\mathcal{M}_f(E)$ corresponding to the transition kernels $Q_t(\mu, d\nu)$ is called the (ξ, ψ)-*superprocess*. By specializing the key formula (10) to constant functions, one easily sees that the "total mass process" $\langle Z, 1 \rangle$ is a ψ-CSBP. When ξ is Brownian motion in $\mathbb{R}^d$ and $\psi(u) = \beta u^2$ (quadratic branching mechanism), the process Z is called *super-Brownian motion*.

Proof. Consider the Markov process Z^ε in the case when its initial value Z_0^ε is distributed according to the law of the Poisson point measure on E with intensity $m_\varepsilon \mu$. By the exponential formula for Poisson measures, we have for $g \in \mathcal{B}_{b+}(E)$,

$$E\big[e^{-\langle m_\varepsilon^{-1} Z_t^\varepsilon, g \rangle}\big] = E\Big[\exp\big(\int Z_0^\varepsilon(dx) \log E_{\delta_x}^\varepsilon \big(e^{-\langle m_\varepsilon^{-1} Z_t^\varepsilon, g \rangle}\big)\big)\Big]$$

$$= \exp\Big(-m_\varepsilon \int \mu(dx)\big(1 - E_{\delta_x}^\varepsilon \big(e^{-\langle m_\varepsilon^{-1} Z_t^\varepsilon, g \rangle}\big)\big)\Big)$$

$$= \exp(-\langle \mu, u_t^\varepsilon \rangle) .$$

From Lemma 4 we get

$$\lim_{\varepsilon \downarrow 0} E\big(e^{-\langle m_\varepsilon^{-1} Z_t^\varepsilon, g \rangle}\big) = \exp(-\langle \mu, u_t \rangle) .$$

Furthermore we see from the proof of Lemma 4 that the convergence is uniform when g varies in the set $\{g \in \mathcal{B}_{b+}(E), 0 \leq g \leq \lambda\} =: \mathcal{H}_\lambda$.

Lemma 6. *Suppose that $R_n(d\nu)$ is a sequence of probability measures on $\mathcal{M}_f(E)$ such that, for every $g \in \mathcal{B}_{b+}(E)$,*

$$\lim_{n \to \infty} \int R_n(d\nu)e^{-\langle \nu, g \rangle} = L(g)$$

with a convergence uniform on the sets $\mathcal{H}_\lambda$. *Then there exists a probability measure* $R(d\nu)$ *on* $\mathcal{M}_f(E)$ *such that*

$$\int R(d\nu)e^{-\langle \nu,g\rangle} = L(g)$$

for every $g \in \mathcal{B}_{b+}(E)$.

We postpone the proof of Lemma 6 and complete the proof of Theorem 5. The first assertion is a consequence of Lemma 6 and the beginning of the proof. The uniqueness of $Q_t(\mu, d\nu)$ follows from the fact that a probability measure $R(d\nu)$ on $\mathcal{M}_f(E)$ is determined by the quantities $\int R(d\nu)\exp(-\langle \nu, g\rangle)$ for $g \in \mathcal{B}_{b+}(E)$ (or even $g \in C_{b+}(E)$). To see this, use standard monotone class arguments to verify that the closure under bounded pointwise convergence of the subspace of $\mathcal{B}_b(\mathcal{M}_f(E))$ generated by the functions $\nu \to \exp(-\langle \nu, g\rangle)$, $g \in \mathcal{B}_{b+}(E)$, is $\mathcal{B}_b(\mathcal{M}_f(E))$.

For the sake of clarity, write $u_t^{(g)}$ for the solution of (9) corresponding to the function g. The Chapman-Kolmogorov equation $Q_{t+s} = Q_t Q_s$ follows from the identity

$$u_t^{(u_s^{(g)})} = u_{t+s}^{(g)},$$

which is easily checked from (9). The measurability of the family of kernels $Q_t(\mu, \cdot)$ is a consequence of (10) and the measurability of the functions $u_t(x)$. Finally, the additivity property follows from (10).   □

Proof of Lemma 6. By a standard result on Polish spaces (see e.g. Parthasarathy [Pa], Chapter 1) we may assume that E is a Borel subset of a compact metric space K. If $g \in \mathcal{B}_{b+}(K)$ we write $L(g) = L(g|_E)$. Also $\mathcal{M}_f(E)$ can obviously be viewed as a subset of $\mathcal{M}_f(K)$. Denote by $\mathcal{M}_1(K)$ the set of all probability measures on K, and fix $\mu_0 \in \mathcal{M}_1(K)$. Consider then the one-to-one mapping

$$\mathcal{J} : \mathcal{M}_f(K) \to [0, \infty] \times \mathcal{M}_1(K)$$

defined by $\mathcal{J}(\nu) = (\langle \nu, 1\rangle, \frac{\nu}{\langle \nu, 1\rangle})$ if $\nu \neq 0$ and $\mathcal{J}(0) = (0, \mu_0)$.

Let $\tilde{R}_n$ denote the image of R_n under $\mathcal{J}$. Then $\tilde{R}_n \in \mathcal{M}_1([0, \infty] \times \mathcal{M}_1(K))$, which is a compact metric space. Hence there exists a subsequence $\tilde{R}_{n_k}$ which converges weakly to $\tilde{R} \in \mathcal{M}_1([0, \infty] \times \mathcal{M}_1(K))$. By our assumption, for every $\varepsilon > 0$,

$$\int \tilde{R}_n(d\ell d\mu)e^{-\varepsilon\ell} = \int R_n(d\nu)e^{-\varepsilon\langle \nu,1\rangle} \xrightarrow[n\to\infty]{} L(\varepsilon) \qquad (11)$$

and because the function $(\ell, \mu) \to e^{-\varepsilon \ell}$ is bounded and continuous on $[0, \infty] \times \mathcal{M}_1(K)$ we get

$$\int \tilde{R}(d\ell d\mu) e^{-\varepsilon \ell} = L(\varepsilon) \ .$$

On the other hand we have also for every n

$$\int \tilde{R}_n (d\ell d\mu) e^{-\varepsilon \ell} = \int R_n(d\nu) e^{-\varepsilon \langle \nu, 1 \rangle} \xrightarrow[\varepsilon \downarrow 0]{} 1 \ . \tag{12}$$

Since the convergence (11) is uniform when ε varies over $(0, 1)$, we easily conclude from (11) and (12) that

$$\lim_{\varepsilon \downarrow 0} \int \tilde{R}(d\ell d\mu) e^{-\varepsilon \ell} = \lim_{\varepsilon \downarrow 0} L(\varepsilon) = 1 \ .$$

It follows that $\tilde{R}$ is in fact supported on $[0, \infty) \times \mathcal{M}_1(K)$.

Let $R(d\nu)$ be the image of $\tilde{R}(d\ell d\mu)$ under the mapping $(\ell, \mu) \to \ell\mu$. For $g \in C_{b+}(K)$,

$$\int R(d\nu) e^{-\langle \mu, g \rangle} = \int \tilde{R}(d\ell d\mu) e^{-\ell \langle \mu, g \rangle} = \lim_{k \to \infty} \int \tilde{R}_{n_k}(d\ell d\mu) e^{-\ell \langle \mu, g \rangle}$$

$$= \lim_{k \to \infty} \int R_{n_k}(d\nu) e^{-\langle \nu, g \rangle} = L(g) \ .$$

(Note that the function $(\ell, \mu) \to e^{-\ell \langle \mu, g \rangle}$ is continuous on $[0, \infty) \times \mathcal{M}_1(K)$.) The uniform convergence assumption ensures that the mapping $g \to L(g)$ is continuous under bounded pointwise convergence: If $g_n \in \mathcal{B}_{b+}(K)$, $g_n \leq C$ and $g_n \to g$ pointwise, then $L(g_n) \to L(g)$. Thus, the set

$$\{g \in \mathcal{B}_{b+}(K), \int R(d\nu) e^{-\langle \nu, g \rangle} = L(g)\}$$

contains $C_{b+}(K)$ and is stable under bounded pointwise convergence. By a standard lemma (see e.g. [EK], p. 111) this set must be $\mathcal{B}_{b+}(K)$.

Finally, by taking $g = 1_{K \setminus E}$ in the equality $\int R(d\nu) e^{-\langle \nu, g \rangle} = L(g)$ we see that, $R(d\nu)$ a.e., ν is supported on E. Therefore we can also view R as a probability measure on $\mathcal{M}_f(E)$. $\qquad\square$

We will write $Z = (Z_t, t \geq 0)$ for the (ξ, ψ)-superprocess whose existence follows from Theorem 6. For $\mu \in \mathcal{M}_f(E)$, we denote by $\mathbb{P}_\mu$ the probability measure under which Z starts at μ. From an intuitive point of view, the measure Z_t should be interpreted as uniformly spread over a cloud of infinitesimal particles moving independently in E according to the spatial motion ξ, and subject continuously to a branching phenomenon governed by ψ.

Remark. For many applications it is important to consider, in addition to the (ξ, ψ)-superprocess Z, the associated historical superprocess which is defined as follows. We replace ξ by the process

$$\tilde{\xi}_t = (\xi_s, 0 \leq s \leq t) \,.$$

Then $\tilde{\xi}_t$ is a Markov process with values in the space $\mathcal{W}$ of finite (càdlàg) paths in E. Note that $\mathcal{W}$ is again a Polish space for an appropriate choice of a distance (based on the Skorokhod metric on càdlàg functions) and that E can be viewed as a subset of $\mathcal{W}$, by identifying a point x with a trivial path with length 0.

The historical superprocess $\tilde{Z}$ is then simply the $(\tilde{\xi}, \psi)$-superprocess. If we start from $\tilde{Z}$ started at $\tilde{Z}_0 = \mu$, for $\mu \in \mathcal{M}_f(E) \subset \mathcal{M}_f(\mathcal{W})$, we can reconstruct a (ξ, ψ)-superprocess Z started at μ via the formula

$$\langle Z_t, f \rangle = \int \tilde{Z}_t(dw) f\big(w(t)\big) \,.$$

This identification is immediate from the formula for the Laplace functional of the transition kernels.

Informally, if we view Z_t as supported on the positions at time t of a set of "infinitesimal particles", $\tilde{Z}_t$ is the corresponding measure on the set of "historical paths" of these particles. We refer to the monograph [DP] for various applications of historical superprocesses.

3   Some properties of superprocesses

Theorem 6 gives no information about the regularity of the sample paths of Z. Such regularity questions are difficult in our general setting (Fitzsimmons [Fi1],[Fi2] gives fairly complete answers). In the present section, we will use elementary methods to derive some weak information on the sample paths of Z. We will rely on the integral equation (9) and certain extensions of this formula that are presented below.

It will be convenient to denote by $\Pi_{s,x}$ the probability measure under which the spatial motion ξ starts from x at time s. Under $\Pi_{s,x}$, ξ_t is only defined for $t \geq s$. We will make the convention that $\Pi_{s,x}\big(f(\xi_t)\big) = 0$ if $t < s$.

Proposition 7. *Let $0 \leq t_1 < t_2 < \cdots < t_p$ and let $f_1, \ldots, f_p \in \mathcal{B}_{b+}(E)$. Then,*

$$\mathbb{E}_\mu \left(\exp - \sum_{i=1}^p \langle Z_{t_i}, f_i \rangle \right) = \exp(-\langle \mu, w_0 \rangle)$$

where the function $(w_t(x), t \geq 0, x \in E)$ is the unique nonnegative solution of the integral equation

$$w_t(x) + \Pi_{t,x} \left(\int_t^\infty \psi(w_s(\xi_s)) \, ds \right) = \Pi_{t,x} \left(\sum_{i=1}^p f_i(\xi_{t_i}) \right). \tag{13}$$

Note that, since ψ is nonnegative, formula (13) and the previous convention imply that $w_t(x) = 0$ for $t > t_p$.

Proof. We argue by induction on p. When $p = 1$, (13) is merely a rewriting of (9): Let $u_t(x)$ be the solution of (9) with $g = f_1$, then

$$w_t(x) = 1_{\{t \leq t_1\}} u_{t_1 - t}(x)$$

solves (13) and

$$\mathbb{E}_\mu(\exp - \langle Z_{t_1}, f_1 \rangle) = \exp - \langle \mu, u_{t_1} \rangle = \exp - \langle \mu, w_0 \rangle .$$

Let $p \geq 2$ and assume that the result holds up to the order $p - 1$. By the Markov property at time t_1,

$$\mathbb{E}_\mu \left(\exp - \sum_{i=1}^p \langle Z_{t_i}, f_i \rangle \right) = \mathbb{E}_\mu \left(\exp(-\langle Z_{t_1}, f_1 \rangle) \mathbb{E}_{Z_{t_1}} \left(\exp - \sum_{i=2}^p \langle Z_{t_i - t_1}, f_i \rangle \right) \right)$$

$$= \mathbb{E}_\mu(\exp(-\langle Z_{t_1}, f_1 \rangle - \langle Z_{t_1}, \tilde{w}_0 \rangle)$$

where $\tilde{w}$ solves

$$\tilde{w}_t(x) + \Pi_{t,x} \left(\int_t^\infty \psi(\tilde{w}_s(\xi_s)) \, ds \right) = \Pi_{t,x} \left(\sum_{i=2}^p f_i(\xi_{t_i - t_1}) \right).$$

By the case $p = 1$ we get

$$\mathbb{E}_\mu \left(\exp - \sum_{i=1}^p \langle Z_{t_i}, f_i \rangle \right) = \exp - \langle \mu, \bar{w}_0 \rangle ,$$

with

$$\bar{w}_t(x) + \Pi_{t,x} \left(\int_t^\infty \psi(\bar{w}_s(\xi_s)) \, ds \right) = \Pi_{t,x} \left(f_1(\xi_{t_1}) + \tilde{w}_0(\xi_{t_1}) \right).$$

We complete the induction by observing that the function $w_t(x)$ defined by

$$w_t(x) = 1_{\{t \leq t_1\}} \bar{w}_t(x) + 1_{\{t > t_1\}} \tilde{w}_{t - t_1}(x)$$

solves (13).

Finally the uniqueness of the nonnegative solution of (13) easily follows from Gronwall's lemma (note that any nonnegative solution w of (13) is automatically bounded and such that $w_t(x) = 0$ for $t > t_p$). $\qquad\square$

Remark. The same proof shows that

$$\exp -\langle \mu, w_t \rangle = \mathbb{E}_{t,\mu}\left(\exp -\sum_{i=1}^{p}\langle Z_{t_i}, f_i \rangle\right) = \mathbb{E}_\mu\left(\exp -\sum_{\substack{i=1 \\ t_i \geq t}}^{p}\langle Z_{t_i-t}, f_i \rangle\right).$$

From now on, we assume that $t \to \xi_t$ is continuous in probability under Π_x, for every $x \in E$.

Proposition 8. *For every* $\mu \in \mathcal{M}_f(E)$, *the process* $(Z_s, s \geq 0)$ *is continuous in probability under* $\mathbb{P}_\mu$.

Proof. Recall (see e.g. [Pa]) that the topology on $\mathcal{M}_f(E)$ may be defined by a metric of the form

$$d(\mu, \mu') = \sum_{n=0}^{\infty}\left(|\langle \mu, f_n \rangle - \langle \mu', f_n \rangle| \wedge 2^{-n}\right)$$

where (f_n) is a suitably chosen sequence in $C_{b+}(E)$. Because of this observation it is enough to prove that for every $f \in C_{b+}(E)$, $\langle Z_s, f \rangle$ is continuous in probability. Equivalently, we have to prove that for every $r \geq 0$ and $\lambda, \gamma \geq 0$

$$\lim_{r' \to r} \mathbb{E}_\mu\left(\exp(-\lambda\langle Z_r, f \rangle - \gamma\langle Z_{r'}, f \rangle)\right) = \mathbb{E}_\mu\left(\exp -(\lambda + \gamma)\langle Z_r, f \rangle\right).$$

By Proposition 7, the expectations in the previous displayed formula are computed in terms of the functions $w_t^r(x), v_t^{r,r'}(x)$ that solve the integral equations

$$w_t^r(x) + \Pi_{t,x}\left(\int_t^\infty \psi(w_s^r(\xi_s))\,ds\right) = (\lambda + \gamma)\Pi_{t,x}\left(f(\xi_r)\right),$$

$$v_t^{r,r'}(x) + \Pi_{t,x}\left(\int_t^\infty \psi(v_s^{r,r'}(\xi_s))\,ds\right) = \lambda\Pi_{t,x}\left(f(\xi_r)\right) + \gamma\Pi_{t,x}\left(f(\xi_{r'})\right).$$

The proof of Proposition 8 then reduces to checking that $v_t^{r,r'}(x) \to w_t^r(x)$ as $r' \to r$. However, from the previous integral equations we get

$$|v_t^{r,r'}(x) - w_t^r(x)| \leq C\Pi_{t,x}\left(\int_t^{r \vee r'}|v_s^{r,r'}(\xi_s) - w_s^r(\xi_s)|ds\right) + H_t^{r,r'}(x),$$

where $H_t^{r,r'}(x) = \gamma \left| \Pi_{t,x}\left(f(\xi_r) - f(\xi_{r'})\right) \right|$ tends to 0 as $r' \to r$, except possibly for $t = r$, and is also uniformly bounded. By iterating the previous bound, as in the proof of the Gronwall lemma, we easily conclude that $|v_0^{r,r'}(x) - w_0^r(x)|$ goes to 0 as $r' \to r$. $\square$

It follows from Proposition 8 that we can choose a measurable modification of the process Z (meaning that $(t, w) \to Z_t(w)$ is measurable). Precisely, we can find an increasing sequence (D_n) of discrete countable subsets of $\mathbb{R}_+$ such that, if $d_n(t) = \inf\{r > t, r \in D_n\}$, the process

$$Z_t' = \begin{cases} \lim_{n\to\infty} Z_{d_n(t)} & \text{if the limit exists,} \\ 0 & \text{if not,} \end{cases}$$

is such that $Z_t' = Z_t$ a.s. for every $t \geq 0$ (and Z' is clearly measurable).

From now on, we systematically replace Z by Z'. Recall from Section 1 that the total mass process $\langle Z, 1 \rangle$, which is a ψ-CSBP, has a modification with càdlàg paths, hence is a.s. bounded over any bounded countable subset of $\mathbb{R}_+$. From our choice of the measurable modification, it follows that $(\langle Z_t, 1 \rangle, t \in [0, T])$ is a.s. bounded, for any $T > 0$.

The measurability property allows us to consider integrals of the form $\int_0^\infty dt\, h(t)\langle Z_t, f \rangle$, where h and f are nonnegative and measurable on $\mathbb{R}_+$ and E respectively.

Corollary 9. Let $f \in \mathcal{B}_{b+}(E)$ and $h \in \mathcal{B}_{b+}(\mathbb{R}_+)$. Assume that h has compact support. Then

$$\mathbb{E}_\mu\left(\exp - \int_0^\infty dt\, h(t)\langle Z_t, f \rangle\right) = \exp - \langle \mu, w_0 \rangle$$

where w is the unique nonnegative solution of the integral equation

$$w_t(x) + \Pi_{t,x}\left(\int_t^\infty \psi(w_s(\xi_s))\, ds\right) = \Pi_{t,x}\left(\int_t^\infty h(s)f(\xi_s)ds\right). \tag{14}$$

Proof. We first assume that both f and h are continuous. Then, Proposition 8 implies that, for every $K > 0$,

$$\lim_{n\to\infty} \mathbb{E}_\mu\left(\int_0^\infty dt\left(\left|h(t)\langle Z_t, f \rangle - h(n^{-1}[nt])\langle Z_{n^{-1}[nt]}, f \rangle\right| \wedge K\right)\right) = 0.$$

Since the process $\langle Z, 1 \rangle$ is locally bounded, it follows that

$$\int_0^\infty dt\, h(t)\langle Z_t, f \rangle = \lim_{n\to\infty} \frac{1}{n} \sum_{i=0}^\infty h\left(\frac{i}{n}\right)\langle Z_{i/n}, f \rangle,$$

in $\mathbb{P}_\mu$-probability. By Proposition 7,

$$\mathbb{E}_\mu\left(\exp - \frac{1}{n} \sum_{i=0}^\infty h\left(\frac{i}{n}\right)\langle Z_{i/n}, f \rangle\right) = \exp - \langle \mu, w_0^n \rangle \tag{15}$$

where

$$w_t^n(x) + \Pi_{t,x}\left(\int_t^\infty \psi(w_s^n(\xi_s))\, ds\right) = \Pi_{t,x}\left(\frac{1}{n}\sum_{i=0}^\infty h\left(\frac{i}{n}\right)f(\xi_{i/n})\right) =: g_n(t,x)\ .$$

The functions $g_n(t,x)$ are uniformly bounded, and converge pointwise to $g(t,x) = \Pi_{t,x}\left(\int_t^\infty ds\, h(s)f(\xi_s)\right)$. By the remark following the proof of Proposition 7, we also know that

$$w_t^n(x) = -\log \mathbb{E}_{t,\delta_x}\left(\exp - \sum_{i=0}^\infty h\left(\frac{i}{n}\right)\langle Z_{i/n}, f\rangle\right).$$

Hence $w_t(x)$ converges as $n \to \infty$ to

$$w_t(x) = -\log \mathbb{E}_{t,\delta_x}\left(\exp - \int_t^\infty ds\, h(s)\langle Z_s, f\rangle\right).$$

We can then pass to the limit in the integral equation for w^n to get that w satisfies (14) (and is the unique nonnegative solution by Gronwall's lemma). The desired result follows by passing to the limit $n \to \infty$ in (15), in the case when f and h are continuous. In the general case, we use the fact that the property of Corollary 9 is stable under bounded pointwise convergence.   □

4   Calculations of moments

Recall that $\alpha \geq 0$ is the coefficient of the linear part in ψ. To simplify notation we write $T_t f(x) = \Pi_x(f(\xi_t))$ for the semigroup of ξ.

Proposition 10. *For every* $f \in \mathcal{B}_{b+}(E)$, $t \geq 0$,

$$\mathbb{E}_\mu(\langle Z_t, f\rangle) = e^{-\alpha t}\langle \mu, T_t f\rangle\ .$$

Proof. First observe that ψ is differentiable at 0 and $\psi'(0) = \alpha$. Then

$$\mathbb{E}_\mu(\langle Z_t, f\rangle) = \lim_{\lambda \downarrow 0}\uparrow \frac{1 - \mathbb{E}_\mu(\exp -\lambda\langle Z_t, f\rangle)}{\lambda} = \lim_{\lambda \downarrow 0}\uparrow \frac{1 - e^{-\langle \mu, v_t^\lambda\rangle}}{\lambda}$$

where $v_t^\lambda(x) = -\log \mathbb{E}_{\delta_x}(\exp -\lambda\langle X_t, f\rangle)$ solves

$$v_t^\lambda(x) + \Pi_x\left(\int_0^t \psi(v_{t-s}^\lambda(\xi_s))\, ds\right) = \lambda\Pi_x(f(\xi_t))\ .$$

From the Hölder inequality, the function $\lambda \to v_t^\lambda(x)$ is concave. It follows that the limit

$$h_t(x) = \lim_{\lambda \downarrow 0}\uparrow \frac{v_t^\lambda(x)}{\lambda}$$

exists and this limit is obviously bounded above by $\Pi_x(f(\xi_t))$. By passing to the limit in the integral equation for v_t^λ we get

$$h_t(x) + \alpha\Pi_x\left(\int_0^t h_{t-s}(\xi_s)\,ds\right) = \Pi_x(f(\xi_t)) \,.$$

However, it is easy to see that the unique solution to this integral equation is $h_t(x) = e^{-\alpha t}\Pi_x(f(\xi_t))$. The desired result follows. $\square$

Moments of higher order need not exist in general. However, they do exist when $\pi = 0$, in particular in the case of the quadratic branching mechanism $\psi(u) = \beta u^2$. These moments can be computed from our formulas for the Laplace functionals. To illustrate the method we consider the case of second moments in the next proposition.

Proposition 11. *Suppose that* $\psi(u) = \beta u^2$. *Then,*

$$\mathbb{E}_\mu(\langle Z_t, f\rangle^2) = \langle\mu, T_tf\rangle^2 + 2\beta\int_0^t \langle\mu, T_s((T_{t-s}f)^2)\rangle ds$$

and more generally

$$\mathbb{E}_\mu(\langle Z_t, f\rangle\langle Z_{t'}, g\rangle) = \langle\mu, T_tf\rangle\langle\mu, T_{t'}g\rangle + 2\beta\int_0^{t\wedge t'} \langle\mu, T_s((T_{t-s}f)(T_{t'-s}g))\rangle ds \,.$$

Proof. First observe that

$$\mathbb{E}_\mu\left(e^{-\lambda\langle Z_t, f\rangle} - 1 + \lambda\langle Z_t, f\rangle\right) = e^{-\langle\mu, v_t^\lambda\rangle} - 1 + \lambda\langle\mu, T_tf\rangle \tag{16}$$

where

$$v_t^\lambda(x) + \beta\int_0^t T_s((v_{t-s}^\lambda)^2)(x)ds = \lambda T_tf(x) \,.$$

It readily follows from the last equation that as $\lambda \to 0$,

$$v_t^\lambda(x) = \lambda T_tf(x) - \lambda^2\beta\int_0^t T_s((T_{t-s}f)^2)(x)ds + O(\lambda^3)$$

with a remainder uniform in $x \in E$. Since

$$\langle Z_t, f\rangle^2 = \lim_{\lambda\downarrow 0}\frac{2}{\lambda^2}\left(e^{-\lambda\langle Z_t, f\rangle} - 1 + \lambda\langle Z_t, f\rangle\right)$$

it follows from (16) and Fatou's lemma that $\mathbb{E}_\mu(\langle Z_t, f\rangle^2) < \infty$, and then by dominated convergence that

$$\mathbb{E}_\mu(\langle Z_t, f\rangle^2) = \lim_{\lambda\downarrow 0}\frac{2}{\lambda^2}\left(e^{-\langle\mu, v_t^\lambda\rangle} - 1 + \lambda\langle\mu, T_t f\rangle\right)$$

$$= \langle\mu, T_t f\rangle^2 + 2\beta\int_0^t \langle\mu, T_s\big((T_{t-s}f)^2\big)\rangle ds$$

By polarization we easily get

$$\mathbb{E}_\mu\big(\langle Z_t, f\rangle\langle Z_t, g\rangle\big) = \langle\mu, T_t f\rangle\langle\mu, T_t g\rangle + 2\beta\int_0^t \langle\mu, T_s\big((T_{t-s}f)(T_{t-s}g)\big)\rangle ds .$$

Finally, if $t \leq t'$, the Markov property and Proposition 10 give

$$\mathbb{E}_\mu(\langle Z_t, f\rangle\langle Z_t, g\rangle) = \mathbb{E}_\mu\big(\langle Z_t, f\rangle\mathbb{E}_{Z_t}(\langle Z_{t'-t}, g\rangle)\big) = \mathbb{E}_\mu(\langle Z_t, f\rangle\langle Z_t, T_{t'-t}g\rangle) .$$

The desired result follows.   □

Exercise. From Proposition 11 one also easily obtains that

$$\mathbb{E}_\mu\left(\int\int Z_t(dy)Z_t(dy')\varphi(y, y')\right) = \int\int \mu T_t(dy)\mu T_t(dy')\varphi(y, y')$$

$$+ 2\beta\int_0^t ds\int \mu T_s(dz)\int\int T_{t-s}(z, dy)T_{t-s}(z, dy')\varphi(y, y') .$$

Use this to verify that for super-Brownian motion in $\mathbb{R}^d$ ($d \geq 2$), for $t > 0$ and $\varepsilon \in (0, 1)$

$$\mathbb{E}_\mu\left(\int\int \frac{Z_t(dy)Z_t(dy')}{|y - y'|^{2-\varepsilon}}\right) < \infty . \tag{17}$$

[*Hint*: If $p_t(x, y)$ is the Brownian transition density, verify that for a fixed $t > 0$

$$\int_0^t ds\int dz\, p_s(x, z)p_{t-s}(z, y)p_{t-s}(z, y') \leq \begin{cases} C(1 + \log_+\frac{1}{|y-y'|}) & \text{if } d = 2 \\ C(|y - y'|^{2-d} + 1) & \text{if } d \geq 3 \end{cases}$$

where the constant C depends only on t.]

The bound (17) and an application of the classical Frostman lemma imply that

$$\dim \mathrm{supp}(Z_t) \geq 2 \quad \text{a.s. on} \quad \{Z_t \neq 0\} ,$$

where $\dim A$ denotes the Hausdorff dimension of A. We will see later that this lower bound is sharp (it is obviously when $d = 2!$).

Chapter III
The Genealogy of Brownian Excursions

We briefly explained in Chapter I that the genealogical structure of the superprocess with branching mechanism $\psi(u) = \beta u^2$ can be coded by Brownian excursions. Our main goal in this chapter is to explain how one can define random trees associated with a Brownian excursion and to give explicit formulas for the distribution of these random trees. As a corollary of our results, we also recover the finite-dimensional marginals of Aldous' continuum random tree.

1 The Itô excursion measure

We denote by $(B_t, t \geq 0)$ a linear Brownian motion, which starts at x under the probability measure P_x. We also set $T_0 = \inf\{t \geq 0, B_t = 0\}$. For $x > 0$, the density of the law of T_0 under P_x is

$$q_x(t) = \frac{x}{\sqrt{2\pi t^3}}\, e^{-\frac{x^2}{2t}}.$$

A major role in what follows will be played by the Itô measure $n(de)$ of positive excursions. This is an infinite measure on the set E_0 of excursions, that is of continuous mappings $e : \mathbb{R}_+ \longrightarrow \mathbb{R}_+$ such that $e(s) > 0$ if and only if $s \in (0, \sigma)$, for some $\sigma = \sigma(e) > 0$ called the length or duration of the excursion e.

For most of our purposes in this chapter, it will be enough to know that $n(de)$ has the following two (characteristic) properties:

(i) For every $t > 0$, and every measurable function $f : \mathbb{R}_+ \longrightarrow \mathbb{R}_+$ such that $f(0) = 0$,

$$\int n(de)\, f(e(t)) = \int_0^\infty dx\, q_x(t)\, f(x). \tag{1}$$

(ii) Let $t > 0$ and let Φ and Ψ be two nonnegative measurable functions defined respectively on $C([0, t], \mathbb{R}_+)$ and $C(\mathbb{R}_+, \mathbb{R}_+)$. Then,

$$\int n(de)\, \Phi(e(r), 0 \leq r \leq t)\Psi(e(t + r), r \geq 0)$$

$$= \int n(de)\, \Phi(e(r), 0 \leq r \leq t)\, E_{e(t)}\big(\Psi(B_{r \wedge T_0}, r \geq 0)\big).$$

Note that (i) implies $n(\sigma > t) = n(e(t) > 0) = (2\pi t)^{-1/2} < \infty$. Property (ii) means that the process $(e(t), t > 0)$ is Markovian under n with the transition kernels of Brownian motion absorbed at 0.

Let us also recall the useful formula $n\left(\sup_{s\geq 0} e(s) > \varepsilon\right) = (2\varepsilon)^{-1}$ for $\varepsilon > 0$.

Lemma 1. *If $f \in \mathcal{B}_+(\mathbb{R}_+)$ and $f(0) = 0$,*

$$n\left(\int_0^\infty dt\, f(e(t))\right) = \int_0^\infty dx\, f(x).$$

Proof. This is a simple consequence of (1). □

For every $t \geq 0$, we set $I_t = \inf_{0\leq s\leq t} B_s$.

Lemma 2. *If $f \in \mathcal{B}_+(\mathbb{R}^3)$ and $x \geq 0$,*

$$E_x\left(\int_0^{T_0} dt\, f(t, I_t, B_t)\right) = 2\int_0^x dy \int_y^\infty dz \int_0^\infty dt\, q_{x+z-2y}(t)\, f(t, y, z). \quad (2)$$

In particular, if $g \in \mathcal{B}_+(\mathbb{R}^2)$,

$$E_x\left(\int_0^{T_0} dt\, g(I_t, B_t)\right) = 2\int_0^x dy \int_y^\infty dz\, g(y, z). \quad (3)$$

Proof. Since

$$E_x\left(\int_0^{T_0} dt\, f(t, I_t, B_t)\right) = \int_0^\infty dt\, E_x\left(f(t, I_t, B_t)\, 1_{\{I_t > 0\}}\right),$$

the lemma follows from the explicit formula

$$E_x\left(g(I_t, B_t)\right) = \int_{-\infty}^x dy \int_y^\infty dz\, \frac{2(x + z - 2y)}{\sqrt{2\pi t^3}}\, e^{-\frac{(x+z-2y)^2}{2t}}\, g(y, z)$$

which is itself a consequence of the reflection principle for linear Brownian motion. □

2 Binary trees

We will consider ordered rooted binary trees. Such a tree describes the genealogy of a population starting with one ancestor (the root ϕ), where each individual can have 0 or 2 children, and the population becomes extinct after a finite number of generations (the tree is finite). The tree is ordered, which means that we put an order on the two children of each individual.

Formally we may and will define a tree as a finite subset T of $\cup_{n=0}^{\infty}\{1,2\}^n$ (with $\{1,2\}^0 = \{\phi\}$) satisfying the obvious conditions:

(i) $\phi \in T$;

(ii) if $(i_1, \ldots, i_n) \in T$ with $n \geq 1$, then $(i_1, \ldots, i_{n-1}) \in T$;

(iii) if $(i_1, \ldots, i_n) \in T$, then either $(i_1, \ldots, i_n, 1) \in T$ and $(i_1, \ldots, i_n, 2) \in T$, or $(i_1, \ldots, i_n, 1) \notin T$ and $(i_1, \ldots, i_n, 2) \notin T$.

The elements of T are the vertices (or individuals in the branching process terminology) of the tree. Individuals without children are called leaves. If T and T' are two trees, the concatenation of T and T', denoted by $T * T'$, is defined in the obvious way: For $n \geq 1$, $(i_1, \ldots, i_n)$ belongs to $T * T'$ if and only if $i_1 = 1$ and $(i_2, \ldots, i_n)$ belongs to T, or $i_1 = 2$ and $(i_2, \ldots, i_n)$ belongs to T'.

Note that $T * T' \neq T' * T$ in general.

For $p \geq 1$, we denote by $\mathbb{T}_p$ the set of all (ordered rooted binary) trees with p leaves. It is easy to compute $a_p = \operatorname{Card} \mathbb{T}_p$. Obviously $a_1 = 1$ and if $p \geq 2$, decomposing the tree at the root shows that $a_p = \sum_{j=1}^{p-1} a_j a_{p-j}$. It follows that

$$a_p = \frac{1 \times 3 \times \ldots \times (2p-3)}{p!} 2^{p-1}.$$

A *marked tree* is a pair $(T, \{h_v, v \in T\})$, where $h_v \geq 0$ for every $v \in T$. Intuitively, h_v represents the lifetime of individual v.

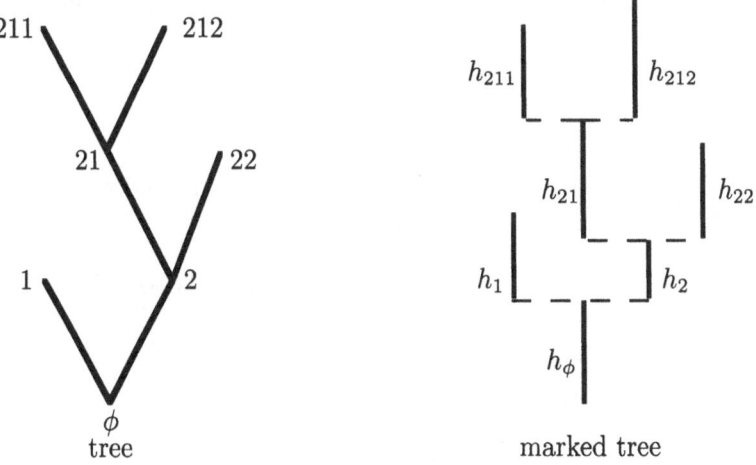

tree   marked tree

Fig. 1

We denote by $\mathcal{T}_p$ the set of all marked trees with p leaves. Let $\theta = (T, \{h_v, v \in T\}) \in \mathcal{T}_p$, $\theta' = (T', \{h'_v, v \in T'\}) \in \mathcal{T}_{p'}$, and $h \geq 0$. The concatenation

$$\theta \underset{h}{*} \theta'$$

is the element of $\mathcal{T}_{p+p'}$ whose "skeleton" is $T * T'$ and such that the marks of vertices in T, respectively in T', become the marks of the corresponding vertices in $T * T'$, and finally the mark of ϕ in $T * T'$ is h.

3 The tree associated with an excursion

Let $e : [a, b] \longrightarrow \mathbb{R}_+$ be a continuous function defined on a subinterval $[a, b]$ of $\mathbb{R}_+$. For every $a \leq u \leq v \leq b$, we set

$$m(u, v) = \inf_{u \leq t \leq v} e(t).$$

Let $t_1, \ldots, t_p \in \mathbb{R}_+$ be such that $a \leq t_1 \leq t_2 \leq \cdots \leq t_p \leq b$. We will now construct a marked tree

$$\theta(e, t_1, \ldots, t_p) = (T(e, t_1, \ldots, t_p), \{h_v(e, t_1, \ldots, t_p), v \in T\}) \in \mathcal{T}_p$$

associated with the function e and the times $t_1, \ldots, t_p$. We proceed by induction on p. If $p = 1$, $T(e, t_1)$ is the unique element of $\mathbb{T}_1$, and $h_\phi(e, t_1) = e(t_1)$. If $p = 2$, $T(e, t_1, t_2)$ is the unique element of $\mathbb{T}_2$, $h_\phi = m(t_1, t_2)$, $h_1 = e(t_1) - m(t_1, t_2)$, $h_2 = e(t_2) - m(t_1, t_2)$.

Then let $p \geq 3$ and suppose that the tree has been constructed up to the order $p - 1$. Let $j = \inf\{i \in \{1, \ldots, p - 1\}, m(t_i, t_{i+1}) = m(t_1, t_p)\}$. Define e' and e'' by the formulas

$$e'(t) = e(t) - m(t_1, t_p), \qquad t \in [t_1, t_j],$$
$$e''(t) = e(t) - m(t_1, t_p), \qquad t \in [t_{j+1}, t_p].$$

By the induction hypothesis, we can associate with e' and $t_1, \ldots, t_j$, respectively with e'' and $t_{j+1}, \ldots, t_p$, a tree

$$\theta(e', t_1, \ldots, t_j) \in \mathcal{T}_j, \quad \text{resp.} \quad \theta(e'', t_{j+1}, \ldots, t_p) \in \mathcal{T}_{p-j}.$$

We set

$$\theta(e, t_1, \ldots, t_p) = \theta(e', t_1, \ldots, t_j) \underset{m(t_1, t_p)}{*} \theta(e'', t_{j+1}, \ldots, t_p).$$

See Fig. I.3 for an example where

$$T(e, t_1, \ldots, t_4) = \{\phi, (1), (2), (1, 1), (1, 2), (2, 1), (2, 2)\}.$$

4   The law of the tree associated with an excursion

Our goal is now to determine the law of the tree $\theta(e, t_1, \ldots, t_p)$ when e is chosen according to the Itô measure of excursions, and $(t_1, \ldots, t_p)$ according to the Lebesgue measure on $[0, \sigma(e)]^p$.

Proposition 3. *For $f \in \mathcal{B}_+(\mathbb{R}_+^{2p-1})$,*

$$n\left(\int_{\{0 \leq t_1 \leq \cdots \leq t_p \leq \sigma\}} dt_1 \ldots dt_p \, f\big(m(t_1, t_2), \ldots, m(t_{p-1}, t_p), e(t_1), \ldots, e(t_p)\big) \right)$$

$$= 2^{p-1} \int_{\mathbb{R}_+^{2p-1}} d\alpha_1 \ldots d\alpha_{p-1} d\beta_1 \ldots d\beta_p$$

$$\times \left(\prod_{i=1}^{p-1} 1_{[0, \beta_i \wedge \beta_{i+1}]}(\alpha_i) \right) f(\alpha_1, \ldots, \alpha_{p-1}, \beta_1, \ldots, \beta_p).$$

Proof. This is a simple consequence of Lemmas 1 and 2. For $p = 1$, the result is exactly Lemma 1. We proceed by induction on p using property (ii) of the Itô measure and then (3):

$$n\left(\int_{\{0 \leq t_1 \leq \cdots \leq t_p \leq \sigma\}} dt_1 \ldots dt_p \, f(m(t_1, t_2), \ldots, m(t_{p-1}, t_p), e(t_1), \ldots, e(t_p)) \right)$$

$$= n\left(\int_{\{0 \leq t_1 \leq \cdots \leq t_{p-1} \leq \sigma\}} dt_1 \ldots dt_{p-1} \right.$$

$$E_{e(t_{p-1})}\left(\int_0^{T_0} dt \, f(m(t_1, t_2), \ldots, m(t_{p-2}, t_{p-1}), I_t, e(t_1), \ldots, e(t_{p-1}), B_t) \right) \Big)$$

$$= 2 n\left(\int_{\{0 \leq t_1 \leq \cdots \leq t_{p-1} \leq \sigma\}} dt_1 \ldots dt_{p-1} \int_0^{e(t_{p-1})} d\alpha_{p-1} \int_{\alpha_{p-1}}^\infty d\beta_p \right.$$

$$\times f(m(t_1, t_2), \ldots, m(t_{p-2}, t_{p-1}), \alpha_{p-1}, e(t_1), \ldots, e(t_{p-1}), \beta_p) \Big).$$

The proof is then completed by using the induction hypothesis.   □

The uniform measure Λ_p on $\mathcal{T}_p$ is defined by

$$\int \Lambda_p(d\theta) \, F(\theta) = \sum_{T \in \mathbb{T}_p} \int \prod_{v \in T} dh_v \, F(T, \{h_v, v \in T\}).$$

Theorem 4. *The law of the tree $\theta(e, t_1, \ldots, t_p)$ under the measure*

$$n(de) \, 1_{\{0 \leq t_1 \leq \cdots \leq t_p \leq \sigma(e)\}} dt_1 \ldots dt_p$$

is $2^{p-1} \Lambda_p$.

Proof. From the construction in Section 3, we have

$$\theta(e, t_1, \ldots, t_p) = \Gamma_p(m(t_1, t_2), \ldots, m(t_{p-1}, t_p), e(t_1), \ldots, e(t_p)),$$

where Γ_p is a measurable function from $\mathbb{R}_+^{2p-1}$ into $\mathcal{T}_p$. Denote by Δ_p the measure on $\mathbb{R}_+^{2p-1}$ defined by

$$\Delta_p(d\alpha_1 \ldots d\alpha_{p-1} d\beta_1 \ldots d\beta_p) = \left(\prod_{i=1}^{p-1} 1_{[0, \beta_i \wedge \beta_{i+1}]}(\alpha_i) \right) d\alpha_1 \ldots d\alpha_{p-1} d\beta_1 \ldots d\beta_p.$$

In view of Proposition 3, the proof of Theorem 4 reduces to checking that $\Gamma_p(\Delta_p) = \Lambda_p$. For $p = 1$, this is obvious.

Let $p \geq 2$ and suppose that the result holds up to order $p - 1$. For every $j \in \{1, \ldots, p-1\}$, let H_j be the subset of $\mathbb{R}_+^{2p-1}$ defined by

$$H_j = \{(\alpha_1, \ldots, \alpha_{p-1}, \beta_1, \ldots, \beta_p); \alpha_j < \alpha_i \text{ for every } i \neq j\}.$$

Then,

$$\Delta_p = \sum_{j=1}^{p-1} 1_{H_j} \cdot \Delta_p.$$

On the other hand, it is immediate to verify that $1_{H_j} \cdot \Delta_p$ is the image of the measure

$$\Delta_j(d\alpha_1' \ldots d\beta_j') \otimes 1_{(0,\infty)}(h) dh \otimes \Delta_{p-j}(d\alpha_1'' \ldots d\beta_{p-j}'')$$

under the mapping $\Phi : (\alpha_1', \ldots, \beta_j', h, \alpha_1'' \ldots, \beta_{p-j}'') \longrightarrow (\alpha_1, \ldots, \beta_p)$ defined by

$$\begin{aligned}
\alpha_j &= h, \\
\alpha_i &= \alpha_i' + h && \text{for } 1 \leq i \leq j - 1, \\
\beta_i &= \beta_i' + h && \text{for } 1 \leq i \leq j, \\
\alpha_i &= \alpha_{i-j}'' + h && \text{for } j+1 \leq i \leq p - 1, \\
\beta_i &= \beta_{i-j}'' + h && \text{for } j+1 \leq i \leq p.
\end{aligned}$$

The construction by induction of the tree $\theta(e, t_1, \ldots, t_p)$ exactly shows that

$$\Gamma_p \circ \Phi(\alpha_1', \ldots, \beta_j', h, \alpha_1'' \ldots, \beta_{p-j}'') = \Gamma_j(\alpha_1', \ldots, \beta_j') \underset{h}{*} \Gamma_{p-j}(\alpha_1'' \ldots, \beta_{p-j}'').$$

Together with the induction hypothesis, the previous observations imply that for any $f \in \mathcal{B}_+(\mathcal{T}_p)$,

$$\begin{aligned}
&\int \Delta_p(du) \, 1_{H_j}(u) \, f(\Gamma_p(u)) \\
&= \int_0^\infty dh \iint \Delta_j(du') \Delta_{p-j}(du'') \, f\left(\Gamma_p(\Phi(u', h, u''))\right) \\
&= \int_0^\infty dh \iint \Delta_j(du') \Delta_{p-j}(du'') \, f\left(\Gamma_j(u') \underset{h}{*} \Gamma_{p-j}(u'')\right) \\
&= \int_0^\infty dh \int \Lambda_j \underset{h}{*} \Lambda_{p-j}(d\theta) \, f(\theta)
\end{aligned}$$

where we write $\Lambda_j \underset{h}{*} \Lambda_{p-j}$ for the image of $\Lambda_j(d\theta)\Lambda_{p-j}(d\theta')$ under the mapping $(\theta, \theta') \longrightarrow \theta \underset{h}{*} \theta'$. To complete the proof, simply note that

$$\Lambda_p = \sum_{j=1}^{p-1} \int_0^\infty dh\, \Lambda_j \underset{h}{*} \Lambda_{p-j}. \qquad \square$$

5 The normalized excursion and Aldous' continuum random tree

In this section, we propose to calculate the law of the tree $\theta(e, t_1, \ldots, t_p)$ when e is chosen according to the law of the Brownian excursion conditioned to have duration 1, and $t_1, \ldots, t_p$ are chosen according to the probability measure $p! 1_{\{0 \le t_1 \le \cdots \le t_p \le 1\}} dt_1 \ldots dt_p$. In contrast with the measure Λ_p of Theorem 4, we get for every p a probability measure on $\mathcal{T}_p$. These probability measures are compatible in a certain sense and they can be identified with the finite-dimensional marginals of Aldous' continuum random tree (this identification is obvious if the CRT is described by the coding explained in Chapter I).

We first recall a few basic facts about the normalized Brownian excursion. There exists a unique collection of probability measures $(n_{(s)}, s > 0)$ on E_0 such that the following properties hold:

(i) For every $s > 0$, $n_{(s)}(\sigma = s) = 1$.

(ii) For every $\lambda > 0$ and $s > 0$, the law under $n_{(s)}(de)$ of $e_\lambda(t) = \sqrt{\lambda} e(t/\lambda)$ is $n_{(\lambda s)}$.

(iii) For every Borel subset A of E_0,

$$n(A) = \frac{1}{2}(2\pi)^{-1/2} \int_0^\infty s^{-3/2}\, n_{(s)}(A)\, ds.$$

The measure $n_{(1)}$ is called the law of the normalized Brownian excursion.

Our first goal is to get a statement more precise than Theorem 4 by considering the pair $(\theta(e, t_1, \ldots, t_p), \sigma)$ instead of $\theta(e, t_1, \ldots, t_p)$. If $\theta = (T, \{h_v, v \in T\})$ is a marked tree, the length of θ is defined in the obvious way by

$$L(\theta) = \sum_{v \in T} h_v.$$

Proposition 5. *The law of the pair $(\theta(e, t_1, \ldots, t_p), \sigma)$ under the measure*

$$n(de)\, 1_{\{0 \le t_1 \le \cdots \le t_p \le \sigma(e)\}} dt_1 \ldots dt_p$$

is

$$2^{p-1} \Lambda_p(d\theta) \, q_{2L(\theta)}(s) ds.$$

Proof. Recall the notation of the proof of Theorem 4. We will verify that, for $f \in \mathcal{B}_+(\mathbb{R}_+^{3p})$,

$$n\left(\int_{\{0 \le t_1 \le \cdots \le t_p \le \sigma\}} dt_1 \ldots dt_p \right.$$

$$\left. f\big(m(t_1, t_2), \ldots, m(t_{p-1}, t_p), e(t_1), \ldots, e(t_p), t_1, t_2 - t_1, \ldots, \sigma - t_p\big) \right)$$

$$= 2^{p-1} \int \Lambda_p(d\alpha_1 \ldots d\alpha_{p-1} d\beta_1 \ldots d\beta_p) \int_{\mathbb{R}_+^{p+1}} ds_1 \ldots ds_{p+1} \tag{4}$$

$$\times \, q_{\beta_1}(s_1) q_{\beta_1 + \beta_2 - 2\alpha_1}(s_2) \ldots q_{\beta_{p-1} + \beta_p - 2\alpha_{p-1}}(s_p) q_{\beta_p}(s_{p+1})$$

$$\times \, f(\alpha_1, \ldots, \alpha_{p-1}, \beta_1, \ldots, \beta_p, s_1, \ldots, s_{p+1}).$$

Suppose that (4) holds. It is easy to check (for instance by induction on p) that

$$2 \, L(\Gamma_p(\alpha_1, \ldots, \alpha_{p-1}, \beta_1, \ldots, \beta_p)) = \beta_1 + \sum_{i=1}^{p-1} (\beta_i + \beta_{i-1} - 2\alpha_i) + \beta_p.$$

Using the convolution identity $q_x * q_y = q_{x+y}$, we get from (4), for $f \in \mathcal{B}_+(\mathbb{R}^{2p})$,

$$n\left(\int_{\{0 \le t_1 \le \cdots \le t_p \le \sigma\}} dt_1 \ldots dt_p \, f\big(m(t_1, t_2), \ldots, m(t_{p-1}, t_p), e(t_1), \ldots, e(t_p), \sigma\big) \right)$$

$$= 2^{p-1} \int \Lambda_p(d\alpha_1 \ldots d\alpha_{p-1} d\beta_1 \ldots d\beta_p) \int_0^\infty dt \, q_{2L(\Gamma_p(\alpha_1, \ldots, \beta_p))}(t) \, f(\alpha_1, \ldots, \beta_p, t).$$

As in the proof of Theorem 4, the statement of Proposition 5 follows from this last identity and the equality $\Gamma_p(\Delta_p) = \Lambda_p$.

It remains to prove (4). The case $p = 1$ is easy: By using property (ii) of the Itô measure, then the definition of the function q_x and finally (1), we get

$$\int n(de) \int_0^\sigma dt \, f(e(t), t, \sigma - t) = \int n(de) \int_0^\sigma dt \, E_{e(t)}\big(f(e(t), t, T_0)\big)$$

$$= \int n(de) \int_0^\sigma dt \int_0^\infty dt' \, q_{e(t)}(t') f(e(t), t, t')$$

$$= \int_0^\infty dx \int_0^\infty dt \, q_x(t) \int_0^\infty dt' \, q_x(t') f(x, t, t').$$

Let $p \geq 2$. Applying the Markov property under n successively at t_p and at t_{p-1}, and then using (2), we obtain

$$n\Bigg(\int_{\{0 \leq t_1 \leq \cdots \leq t_p \leq \sigma\}} dt_1 \ldots dt_p$$
$$\times f\big(m(t_1, t_2), \ldots, m(t_{p-1}, t_p), e(t_1), \ldots, e(t_p), t_1, t_2 - t_1, \ldots, \sigma - t_p\big)\Bigg)$$

$$= n\Bigg(\int_{\{0 \leq t_1 \leq \cdots \leq t_{p-1} \leq \sigma\}} dt_1 \ldots dt_{p-1}\, E_{e(t_{p-1})}\Big(\int_0^{T_0} dt \int_0^\infty ds\, q_{B_t}(s) \Big)$$
$$\times f\big(m(t_1, t_2), \ldots, m(t_{p-2}, t_{p-1}), I_t, e(t_1), \ldots$$
$$\ldots, e(t_{p-1}), B_t, t_1, \ldots, t_{p-1} - t_{p-2}, t, s\big)\Bigg)\Bigg)$$

$$= 2n\Bigg(\int_{\{0 \leq t_1 \leq \cdots \leq t_{p-1} \leq \sigma\}} dt_1 \ldots dt_{p-1}$$
$$\times \int_0^{e(t_{p-1})} dy \int_y^\infty dz \int_0^\infty dt \int_0^\infty ds\, q_{e(t_{p-1})+z-2y}(t) q_z(s)$$
$$\times f\big(m(t_1, t_2), \ldots, m(t_{p-2}, t_{p-1}), y, e(t_1), \ldots$$
$$\ldots, e(t_{p-1}), z, t_1, \ldots, t_{p-1} - t_{p-2}, t, s\big)\Bigg).$$

It is then straightforward to complete the proof by induction on p.   □

We can now state and prove the main result of this section.

Theorem 6. *The law of the tree $\theta(e, t_1, \ldots, t_p)$ under the probability measure*

$$p!\, 1_{\{0 \leq t_1 \leq \cdots \leq t_p \leq 1\}} dt_1 \ldots dt_p\, n_{(1)}(de)$$

is

$$p!\, 2^{p+1}\, L(\theta) \exp\big(-2\, L(\theta)^2 \big) \Lambda_p(d\theta).$$

Proof. We equip $\mathcal{T}_p$ with the obvious product topology. Let $F \in C_{b+}(\mathcal{T}_p)$ and $h \in \mathcal{B}_{b+}(\mathbb{R}_+)$. By Proposition 5,

$$\int n(de)\, h(\sigma) \int_{\{0 \leq t_1 \leq \cdots \leq t_p \leq \sigma\}} dt_1 \ldots dt_p\, F\big(\theta(e, t_1, \ldots, t_p)\big)$$

$$= 2^{p-1} \int_0^\infty ds\, h(s) \int \Lambda_p(d\theta)\, q_{2L(\theta)}(s)\, F(\theta).$$

On the other hand, using the properties of the definition of the measures $n_{(s)}$, we have also

$$\int n(de)\, h(\sigma) \int_{\{0\leq t_1 \leq \cdots \leq t_p \leq \sigma\}} dt_1 \ldots dt_p\, F\big(\theta(e, t_1, \ldots, t_p)\big)$$

$$= \frac{1}{2}(2\pi)^{-1/2} \int_0^\infty ds\, s^{-3/2}\, h(s) \int n_{(s)}(de)$$

$$\times \int_{\{0\leq t_1 \leq \cdots \leq t_p \leq s\}} dt_1 \ldots dt_p\, F\big(\theta(e, t_1, \ldots, t_p)\big).$$

By comparing with the previous identity, we get for a.a. $s > 0$,

$$\int n_{(s)}(de) \int_{\{0\leq t_1 \leq \cdots \leq t_p \leq s\}} dt_1 \ldots dt_p\, F\big(\theta(e, t_1, \ldots, t_p)\big)$$

$$= 2^{p+1} \int \Lambda_p(d\theta)\, L(\theta)\, \exp\Big(-\frac{2L(\theta)^2}{s}\Big)\, F(\theta).$$

Both sides of the previous equality are continuous functions of s (use the scaling property of $n_{(s)}$ for the left side). Thus the equality holds for every $s > 0$, and in particular for $s = 1$. This completes the proof. $\square$

Concluding remarks. If we pick $t_1, \ldots, t_p$ independently according to the Lebesgue measure on $[0, 1]$, we can consider the increasing rearrangement $t_1' \leq t_2' \leq \cdots \leq t_p'$ of $t_1, \ldots, t_p$ and define $\theta(e, t_1, \ldots, t_p) = \theta(e, t_1', \ldots, t_p')$. We can also keep track of the initial ordering and consider the tree $\tilde{\theta}(e, t_1, \ldots, t_p)$ defined as the tree $\theta(e, t_1, \ldots, t_p)$ where leaves are labelled $1, \ldots, p$, the leaf corresponding to time t_i receiving the label i. (This labelling has nothing to do with the ordering of the tree.) Theorem 6 implies that the law of the tree $\tilde{\theta}(e, t_1, \ldots, t_p)$ under the probability measure

$$1_{[0,1]^p}(t_1, \ldots, t_p)dt_1 \ldots dt_p\, n_{(1)}(de)$$

has density

$$2^{p+1} L(\theta)\, \exp(-2L(\theta)^2)$$

with respect to $\tilde{\Lambda}_p(d\theta)$, the uniform measure on the set of labelled marked trees.

We can then "forget" the ordering. Define $\theta^*(e, t_1, \ldots, t_p)$ as the tree $\tilde{\theta}(e, t_1, \ldots, t_p)$ without the order structure. Since there are 2^{p-1} possible orderings for a given labelled tree, we get that the law (under the same measure) of the tree $\theta^*(e, t_1, \ldots, t_p)$ has density

$$2^{2p} L(\theta)\, \exp(-2L(\theta)^2)$$

with respect to $\Lambda_p^*(d\theta)$, the uniform measure on the set of labelled marked unordered trees.

For convenience, replace the excursion e by $2e$ (this simply means that all heights are multiplied by 2). We obtain that the law of the tree $\theta^*(2e, t_1, \ldots, t_p)$ has density

$$L(\theta) \exp(-\frac{L(\theta)^2}{2})$$

with respect to $\Lambda_p^*(d\theta)$. It is remarkable that the previous density (apparently) does not depend on p.

In the previous form, we recognize the finite-dimensional marginals of Aldous' continuum random tree [Al1]. To give a more explicit description, the discrete skeleton $T^*(2e, t_1, \ldots, t_p)$ is distributed uniformly on the set of labelled rooted binary trees with p leaves. (This set has b_p elements, with $b_p = p!\, 2^{-(p-1)} a_p = 1 \times 3 \times \cdots \times (2p-3)$.) Then, conditionally on the discrete skeleton, the heights h_v are distributed with the density

$$b_p \left(\sum h_v \right) \exp \left(- \frac{\left(\sum h_v \right)^2}{2} \right)$$

(verify that this is a probability density on $\mathbb{R}_+^{2p-1}$!).

Chapter IV
The Brownian Snake and
Quadratic Superprocesses

In this chapter, we introduce the path-valued process called the Brownian snake and we use this process to give a new construction of superprocesses with branching mechanism $\psi(u) = \beta u^2$. This construction will be applied to connections with partial differential equations in the forthcoming chapters. The proof of the relationship between the Brownian snake and superprocesses relies on our study of the genealogy of Brownian excursions in the previous chapter. In the last sections, under stronger continuity assumptions on the spatial motion, we use the Brownian snake approach to derive various properties of superprocesses.

1 The Brownian snake

As in Chapter II, we consider a Markov process (ξ_t, Π_x) with càdlàg paths and values in a Polish space E and we denote by $\delta(x, y)$ the distance on E. For technical convenience, we will assume a little more than the continuity in probability of ξ under Π_x for every $x \in E$. Precisely, we assume that for every $\varepsilon > 0$

$$\lim_{t \to 0} \left(\sup_{x \in E} \Pi_x \left(\sup_{r \leq t} \delta(x, \xi_r) > \varepsilon \right) \right) = 0. \tag{1}$$

Without the supremum in x, this is simply the right-continuity of paths. Here we require uniformity in x.

Let us introduce the space of finite paths in E. If I is an interval of $\mathbb{R}_+$, we denote by $\mathbb{D}(I, E)$ the Skorokhod space of càdlàg mappings from I into E. We then set

$$\mathcal{W} = \bigcup_{t \geq 0} \mathbb{D}([0, t], E)$$

and if $w \in \mathcal{W}$, we write $\zeta_w = t$ if $w \in \mathbb{D}([0, t], E)$ (ζ_w is called the lifetime of w). We also use the notation $\hat{w} = w(\zeta_w)$ for the terminal point of w. The set $\mathcal{W}$ is equipped with the distance

$$d(w, w') = |\zeta_w - \zeta_{w'}| + d_0\left(w(. \wedge \zeta_w), w'(. \wedge \zeta_{w'})\right),$$

where d_0 is a distance defining the Skorokhod topology on $\mathbb{D}([0, \infty), E)$. It is easy to verify that $(\mathcal{W}, d)$ is a Polish space. It will be convenient to view E as a subset of $\mathcal{W}$, by identifying a point $x \in E$ with the trivial path with initial point x and lifetime $\zeta = 0$. For $x \in E$, we denote by $\mathcal{W}_x$ the set $\{w \in \mathcal{W}, w(0) = x\}$.

Let $w \in \mathcal{W}$ and $a \in [0, \zeta_w]$, $b \geq a$. We define a probability measure $R_{a,b}(w, dw')$ on $\mathcal{W}$ by the following prescriptions:

(i) $\zeta_{w'} = b$, $R_{a,b}(w, dw')$ a.s.

(ii) $w'(t) = w(t)$, for every $t \in [0, a]$, $R_{a,b}(w, dw')$ a.s.

(iii) The law under $R_{a,b}(w, dw')$ of $(w'(a+t), 0 \leq t \leq b-a)$ is the law of $(\xi_t, 0 \leq t \leq b-a)$ under $\Pi_{w(a)}$.

Informally, the path w' is obtained by first restricting w to the time interval $[0, a]$ and then extending the restricted path to $[0, b]$ by using the law of ξ between a and b.

Let $(\beta_s, s \geq 0)$ be a reflected linear Brownian motion (the modulus of a standard linear Brownian motion) started at x. For every $s > 0$, we denote by $\gamma_s^x(da\, db)$ the joint distribution of the pair $(\inf_{0 \leq r \leq s} \beta_r, \beta_s)$. The reflection principle easily gives the explicit form of $\gamma_s^x(da\, db)$:

$$\gamma_s^x(da\, db) = \frac{2(x+b-2a)}{(2\pi s^3)^{1/2}} \exp{-\frac{(x+b-2a)^2}{2s}} 1_{(0<a<b\wedge x)}\, da\, db$$

$$+ 2(2\pi s)^{-1/2} \exp{-\frac{(x+b)^2}{2s}} 1_{(0<b)} \delta_0(da)\, db.$$

Definition. *The ξ-Brownian snake is the Markov process in $\mathcal{W}$, denoted by $(W_s, s \geq 0)$, whose transition kernels Q_s are given by the formula*

$$Q_s(w, dw') = \int\int \gamma_s^{\zeta_w}(da\, db)\, R_{a,b}(w, dw').$$

We will use the notation $\zeta_s = \zeta_{W_s}$ for the lifetime of W_s.

Although the formula for Q_s looks complicated, the behavior of the process W can be described in a simple way. Informally, W_s is a path of ξ started at x, with a random lifetime ζ_s evolving like reflected linear Brownian motion.

When ζ_s decreases the path W_s is simply erased (or shortened) from its tip, and when ζ_s increases the path is extended using the law of ξ for the extension. From the formula for Q_s, it is also clear that $W_s(0) = W_0(0)$ a.s., so that the process W started at w_0 indeed takes values in $\mathcal{W}_x$ with $x = w_0(0)$.

It is maybe not immediate that the collection of kernels Q_s forms a semigroup of transition kernels. (The previous intuitive interpretation should make this property obvious.) We will leave the easy verification of this fact to the reader, and rather give a construction of W that we will use in the remainder of this chapter.

This construction involves defining the conditional distributions of W given the "lifetime process" $(\zeta_s, s \geq 0)$. Fix a starting point $w_0 \in \mathcal{W}$ and set $\zeta_0 = \zeta_{w_0}$. The process W started at w_0 will be constructed on the canonical space $C(\mathbb{R}_+, \mathbb{R}_+) \times \mathcal{W}^{\mathbb{R}_+}$. To this end, denote by P_{ζ_0} the law on $C(\mathbb{R}_+, \mathbb{R}_+)$ of reflected Brownian motion started at ζ_0. Then, for $f \in C(\mathbb{R}_+, \mathbb{R}_+)$ such that $f(0) = \zeta_0$, let $\Theta^f_{w_0}(dw)$ be the law on $\mathcal{W}^{\mathbb{R}_+}$ of the time-inhomogeneous Markov process in $\mathcal{W}$ started at w_0 and whose transition kernel between times s and s' is

$$R_{m(s,s'),f(s')}(w, dw'),$$

where $m(s, s') = \inf_{s \leq r \leq s'} f(s)$. Note that the existence of $\Theta^f_{w_0}$ is an easy application of the Kolmogorov extension theorem. Furthermore, if A is a measurable subset of $\mathcal{W}^{\mathbb{R}_+}$ depending on finitely many coordinates, it is straightforward to verify that the mapping $f \longrightarrow \Theta^f_{w_0}(A)$ is measurable. Thus it makes sense to consider the probability measure

$$\mathbb{P}_{w_0}(df\,dw) = P_{\zeta_0}(df)\Theta^f_{w_0}(dw).$$

The process W started at w_0 is defined under $\mathbb{P}_{w_0}(df\,dw)$ by $W_s(f, w) = w(s)$.

A straightforward calculation of finite-dimensional marginals shows that $(W_s, s \geq 0)$ is under $\mathbb{P}_{w_0}$ a (time-homogeneous) Markov process with transition kernels Q_s. Note that we have $\zeta_s(f, w) = f(s)$, $\mathbb{P}_{w_0}$ a.s., so that the lifetime process is a reflected Brownian motion. (This fact can also be deduced from the form of Q_s.)

It is easy to verify that the kernels Q_s are symmetric with respect to the (invariant) measure

$$M(dw) = \int_0^\infty da\,\Pi^a_x(dw),$$

where $\Pi^a_x(dw) = R_{0,a}(x, dw)$ is the law of the process ξ started at x and stopped at time a. We can thus apply to W the tools of the theory of symmetric Markov processes. We will not give such applications here, but refer the interested reader to [L4] or [L7].

A major role in what follows will be played by the excursion measures of the Brownian snake. For $x \in E$, the excursion measure $\mathbb{N}_x$ is the σ-finite measure defined by

$$\mathbb{N}_x(df\, d\omega) = n(df)\Theta_x^f(d\omega),$$

where $n(df)$ is the Itô excursion measure as in Chapter III. We will see later (under additional regularity assumptions) that the law of $(W_s, s \geq 0)$ under $\mathbb{N}_x$ is indeed the excursion measure of the Brownian snake away from the trivial path x, in the sense of excursion theory for Markov processes. The process W_s can be described informally under $\mathbb{N}_x$ in the same way as under $\mathbb{P}_{w_0}$, with the only difference that the lifetime process ζ_s is now distributed according to the Itô excursion measure.

Lemma 1.
  (i) *For every $\varepsilon > 0$ and $\alpha > 0$,*

$$\lim_{r\downarrow 0} \left(\sup_{s \geq \alpha} \mathbb{P}_{w_0}(d(W_s, W_{s+r}) > \varepsilon) \right) = 0$$

  and the convergence is uniform in $w_0 \in \mathcal{W}$.

  (ii) *Let $f \in C(\mathbb{R}_+, \mathbb{R}_+)$ with compact support and such that $f(0) = 0$. Then, for every $\varepsilon > 0$,*

$$\lim_{r\downarrow 0} \left(\sup_{s \geq 0} \Theta_x^f(d(W_s, W_{s+r}) > \varepsilon) \right) = 0.$$

  The convergence is uniform in $x \in E$, and its rate only depends on a modulus of continuity for f.

Remark. The first part of the lemma implies that the process $(W_s, s \geq 0)$ is continuous in probability under $\mathbb{P}_{w_0}$, except possibly at $s = 0$. On the other hand, it is easy to see that if w_0 has a jump at ζ_{w_0}, then W_s will not be continuous in probability at $s = 0$ under $\mathbb{P}_{w_0}$.

Proof. We prove (i). Let $0 < s < s'$ and let $f \in C(\mathbb{R}_+, \mathbb{R}_+)$ be such that $m(0, s) \leq m(s, s')$. Then, under $\Theta_{w_0}^f$, W_s and $W_{s'}$ are two random paths with respective lifetimes $f(s)$ and $f(s')$, which coincide up to time $m(s, s')$ and then behave independently according to the law of ξ. Let $\varepsilon > 0$ and $\eta > 0$. We can easily bound

$$\mathbb{P}_{w_0}\left(\sup_{t \geq 0} \delta(W_s(t \wedge \zeta_s), W_{s'}(t \wedge \zeta_{s'})) > 2\varepsilon \right)$$

$$\leq P_{\zeta_{w_0}}(m(s,s') < m(0,s)) + P_{\zeta_{w_0}}(\zeta_s - m(s,s') > \eta) + P_{\zeta_{w_0}}(\zeta_{s'} - m(s,s') > \eta)$$

$$+ 2E_{\zeta_{w_0}}\left(1_{\{m(0,s) \leq m(s,s')\}} \Pi_{w_0(m(0,s))}\left(\Pi_{\xi_{m(s,s')-m(0,s)}}\left(\sup_{0 \leq r \leq \eta} \delta(\xi_0, \xi_r) > \varepsilon \right) \right) \right).$$

For any fixed $\eta > 0$, the first three terms of the right side will be small provided that $s' - s$ is small enough and $s \geq \alpha > 0$. On the other hand, the last term goes to 0 as $\eta \to 0$, uniformly in w_0, thanks to our assumption (1). This completes the proof of (i). The argument for (ii) is similar.   □

As we did in Chapter II for superprocesses, we may use the previous lemma to construct a measurable modification of W. We can choose an increasing sequence (D_n) of discrete countable subsets of $\mathbb{R}_+$, with union dense in $\mathbb{R}_+$, in such a way that the following properties hold. If $d_n(s) = \inf\{r \geq s, r \in D_n\}$, the process

$$W'_s = \begin{cases} \lim_{n\to\infty} W_{d_n(s)} & \text{if the limit exists,} \\ W_0 & \text{if not,} \end{cases}$$

satisfies both $\mathbb{P}_{w_0}(W'_s \neq W_s) = 0$ for every $s \geq 0$, $w_0 \in W$ and $\Theta^f_x(W'_s \neq W_s) = 0$ for every $s \geq 0$ and $x \in E$, $n(df)$ a.e. From now on we deal only with this modification and systematically replace W by W'. Note that W' is also a modification of W under $\mathbb{N}_x$, for every $x \in E$.

As in Chapter III, we write $\sigma = \sigma(f)$ under $\mathbb{N}_x$.

2   Finite-dimensional marginals of the Brownian snake

In this section, we briefly derive a description of the finite-dimensional marginals of the Brownian snake, in terms of the marked trees that were introduced in the previous chapter.

Let $\theta \in \mathcal{T}_p$ be a marked tree with p branches. We associate with θ a probability measure on $(\mathcal{W}_x)^p$ denoted by Π^θ_x, which is defined inductively as follows.

If $p = 1$, then $\theta = (\{\phi\}, h)$ for some $h \geq 0$ and we let $\Pi^\theta_x = \Pi^h_x$ be the law of $(\xi_t, 0 \leq t \leq h)$ under Π_x.

If $p \geq 2$, then we can write in a unique way

$$\theta = \theta' \underset{h}{*} \theta'' \,,$$

where $\theta' \in \mathcal{T}_j$, $\theta'' \in \mathcal{T}_{p-j}$, and $j \in \{1, \ldots, p - 1\}$. We then define Π^θ_x by

$$\int \Pi^\theta_x(dw_1, \ldots, dw_p) F(w_1, \ldots, w_p)$$

$$= \Pi_x \left(\int\!\!\int \Pi^{\theta'}_{\xi_h}(dw'_1, \ldots, dw'_j) \Pi^{\theta''}_{\xi_h}(dw''_1, \ldots, dw''_{p-j}) \right.$$

$$\left. F(\xi_{[0,h]} \odot w'_1, \ldots, \xi_{[0,h]} \odot w'_j, \xi_{[0,h]} \odot w''_1, \ldots, \xi_{[0,h]} \odot w''_{p-j}) \right)$$

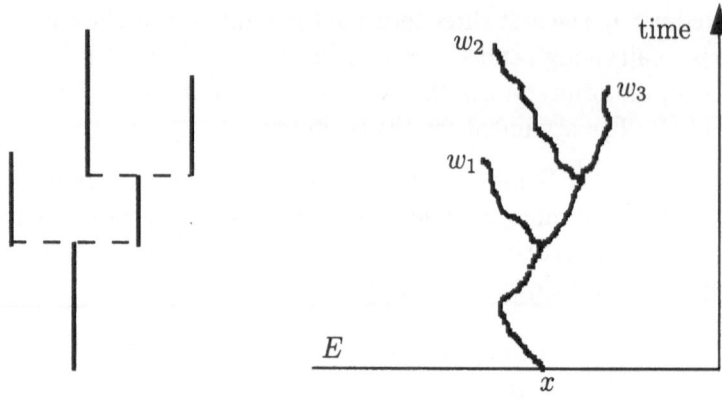

Fig. 1

where $\xi_{[0,h]} \odot w$ denotes the concatenation (defined in an obvious way) of the paths $(\xi_t, 0 \le t \le h)$ and $(w(t), 0 \le t \le \zeta_w)$.

Informally, Π_x^θ is obtained by running independent copies of ξ along the branches of the tree θ. See Fig. 1 for a simple example with $p = 3$.

Proposition 2.

(i) *Let $f \in C(\mathbb{R}_+, \mathbb{R}_+)$ such that $f(0) = 0$, and let $0 \le t_1 \le t_2 \le \cdots \le t_p$. Then the law under Θ_x^f of $(w(t_1), \ldots, w(t_p))$ is $\Pi_x^{\theta(f,t_1,\ldots,t_p)}$.*

(ii) *For any $F \in \mathcal{B}_+(\mathcal{W}_x^p)$,*

$$\mathbb{N}_x \left(\int_{\{0 \le t_1 \le \cdots \le t_p \le \sigma\}} dt_1 \ldots dt_p \, F(W_{t_1}, \ldots, W_{t_p}) \right) = 2^{p-1} \int \Lambda_p(d\theta) \Pi_x^\theta(F).$$

Proof. Assertion (i) follows easily from the definition of Θ_x^f and the construction of the trees $\theta(f, t_1, \ldots, t_p)$. A precise argument can be given using induction on p, but we leave details to the reader. To get (ii), we write

$$\mathbb{N}_x \left(\int_{\{0 \le t_1 \le \cdots \le t_p \le \sigma\}} dt_1 \ldots dt_p \, F(W_{t_1}, \ldots, W_{t_p}) \right)$$

$$= \int n(df) \int_{\{0 \le t_1 \le \cdots \le t_p \le \sigma\}} dt_1 \ldots dt_p \, \Theta_x^f \left(F(W_{t_1}, \ldots, W_{t_p}) \right)$$

$$= \int n(df) \int_{\{0 \le t_1 \le \cdots \le t_p \le \sigma\}} dt_1 \ldots dt_p \, \Pi_x^{\theta(f,t_1,\ldots,t_p)}(F)$$

$$= 2^{p-1} \int \Lambda_p(d\theta) \, \Pi_x^\theta(F).$$

The first equality is the definition of $\mathbb{N}_x$, the second one is part (i) of the proposition, and the last one is Theorem III.4.   □

The cases $p = 1$ and $p = 2$ of Proposition 2 (ii) will be used several times in what follows. Let us rewrite the corresponding formulas in a special case. Recall the notation $\hat{w}$ for the terminal point of w. For any $g \in \mathcal{B}_+(E)$, we have

$$\mathbb{N}_x\left(\int_0^\sigma ds\, g(\hat{W}_s)\right) = \Pi_x\left(\int_0^\infty dt\, g(\xi_t)\right),$$

and

$$\mathbb{N}_x\left(\left(\int_0^\sigma ds\, g(\hat{W}_s)\right)^2\right) = 4\,\Pi_x\left(\int_0^\infty dt\left(\Pi_{\xi_t}\left(\int_0^\infty dr\, g(\xi_r)\right)\right)^2\right).$$

These formulas are reminiscent of the moment formulas for superprocesses obtained in Chapter II in the quadratic branching case. We will see in the next section that this analogy is not a coincidence.

3   The connection with superprocesses

We start with a key technical result.

Proposition 3. *Let $g \in \mathcal{B}_{b+}(\mathbb{R}_+ \times E)$ such that $g(t,y) = 0$ for $t \geq A > 0$. Then the function*

$$u_t(x) = \mathbb{N}_x\left(1 - \exp - \int_0^\sigma ds\, g(t + \zeta_s, \hat{W}_s)\right)$$

solves the integral equation

$$u_t(x) + 2\,\Pi_{t,x}\left(\int_t^\infty dr\,(u_r(\xi_r))^2\right) = \Pi_{t,x}\left(\int_t^\infty dr\, g(r,\xi_r)\right) \tag{1}$$

(recall that the process ξ starts from x at time t under the probability measure $\Pi_{t,x}$).

Proof. For every integer $p \geq 1$, set

$$T^p g(t,x) = \frac{1}{p!}\mathbb{N}_x\left(\left(\int_0^\sigma ds\, g(t + \zeta_s, \hat{W}_s)\right)^p\right).$$

By the case $p = 1$ of Proposition 2 (ii), we have

$$T^1 g(t,x) = \Pi_x\left(\int_0^\infty dr\, g(t + r, \xi_r)\right) \tag{2}$$

Then let $p \geq 2$. Using Proposition 2 (ii) again we have

$$T^p g(t, x) = \mathbb{N}_x \left(\int_{\{0 \leq s_1 \leq \cdots \leq s_p \leq \sigma\}} ds_1 \ldots ds_p \prod_{i=1}^{p} g(t + \zeta_{s_i}, \hat{W}_{s_i}) \right)$$

$$= 2^{p-1} \int \Lambda_p(d\theta) \int \Pi_x^{\theta}(dw_1 \ldots dw_p) \prod_{i=1}^{p} g(t + \zeta_{w_i}, \hat{w}_i)$$

$$= 2^{p-1} \sum_{j=1}^{p-1} \int_0^{\infty} dh \int\!\!\int \Lambda_j(d\theta') \Lambda_{p-j}(d\theta'')$$

$$\Pi_x \left(\left(\int \Pi_{\xi_h}^{\theta'}(dw_1' \cdots dw_j') \prod_{i=1}^{j} g(t + h + \zeta_{w_i'}, \hat{w}_i') \right) \right.$$

$$\left. \times \left(\int \Pi_{\xi_h}^{\theta''}(dw_1'' \cdots dw_{p-j}'') \prod_{i=1}^{p-j} g(t + h + \zeta_{w_i''}, \hat{w}_i'') \right) \right) .$$

In the last equality we used the identity

$$\Lambda_p = \sum_{j=1}^{p-1} \int_0^{\infty} dh \, \Lambda_j \underset{h}{*} \Lambda_{p-j}$$

together with the construction by induction of Π_x^{θ}. We thus get the recursive formula

$$T^p g(t, x) = 2 \sum_{j=1}^{p-1} \Pi_x \left(\int_0^{\infty} dh \, T^j g(t + h, \xi_h) T^{p-j} g(t + h, \xi_h) \right) . \qquad (3)$$

For $p = 1$, (2) gives the bound

$$T^1 g(t, x) \leq C 1_{[0,A]}(t) .$$

Recall from Chapter III the definition of the numbers a_p satisfying $a_p = \sum_{j=1}^{p-1} a_j a_{p-j}$. From the bound for $p = 1$ and (3), we easily get $T^p g(t, x) \leq (2A)^{p-1} C^p a_p \, 1_{[0,A]}(t)$ by induction on p. Hence,

$$T^p g(t, x) \leq (C')^p 1_{[0,A]}(t) .$$

It follows that, for $0 < \lambda < \lambda_0 := (C')^{-1}$,

$$\sum_{p=1}^{\infty} \lambda^p T^p g(t, x) \leq K \, 1_{[0,A]}(t), \qquad (4)$$

for some constant $K < \infty$.

By expanding the exponential we get for $\lambda \in (0, \lambda_0)$

$$u_t^\lambda(x) := \mathbb{N}_x\left(1 - \exp\left(-\lambda \int_0^\sigma ds\, g(t + \zeta_s, \hat{W}_s))\right)\right) = \sum_{p=1}^\infty (-1)^{p+1} \lambda^p T^p g(t, x) .$$

By (3), we have also

$$2\Pi_x\left(\int_0^\infty dr\left(u_{t+r}^\lambda(\xi_r)\right)^2\right)$$

$$= 2\Pi_x\left(\int_0^\infty dr\left(\sum_{p=1}^\infty (-1)^{p+1} \lambda^p T^p g(t + r, \xi_r)\right)^2\right)$$

$$= 2\sum_{p=2}^\infty (-1)^p \lambda^p \sum_{j=1}^{p-1} \Pi_x\left(\int_0^\infty dr\, T^j g(t + r, \xi_r) T^{p-j} g(t + r, \xi_r)\right)$$

$$= \sum_{p=2}^\infty (-1)^p \lambda^p T^p g(t, x) .$$

(The use of Fubini's theorem in the second equality is justified thanks to (4).) From the last equality and the previous formula for $u_t^\lambda(x)$, we get, for $\lambda \in (0, \lambda_0)$,

$$u_t^\lambda(x) + 2\Pi_x\left(\int_0^\infty dr\left(u_{t+r}^\lambda(\xi_r)\right)^2\right) = \lambda T^1 g(t, x) = \lambda \Pi_x\left(\int_0^\infty dr\, g(t + r, \xi_r)\right) .$$

This is the desired integral equation, except that we want it for $\lambda = 1$. Note however that the function $\lambda \longrightarrow u_t^\lambda(x)$ is holomorphic on the domain $\{\text{Re}\,\lambda > 0\}$. Thus, an easy argument of analytic continuation shows that if the previous equation holds for $\lambda \in (0, \lambda_0)$, it must hold for every $\lambda > 0$. This completes the proof. $\qquad\square$

Theorem 4. *Let $\mu \in M_f(E)$ and let*

$$\sum_{i \in I} \delta_{(x_i, f_i, \omega_i)}$$

be a Poisson point measure with intensity $\mu(dx)\mathbb{N}_x(df\,d\omega)$. Write $W_s^i = W_s(f_i, \omega_i)$, $\zeta_s^i = \zeta_s(f_i, \omega_i)$ and $\sigma_i = \sigma(f_i)$ for every $i \in I$, $s \geq 0$. Then there exists a $(\xi, 2u^2)$-superprocess $(Z_t, t \geq 0)$ with $Z_0 = \mu$ such that for every $h \in B_{b+}(\mathbb{R}_+)$ and $g \in B_{b+}(E)$,

$$\int_0^\infty h(t)\langle Z_t, g\rangle dt = \sum_{i \in I} \int_0^{\sigma_i} h(\zeta_s^i) g(\hat{W}_s^i) ds .$$

More precisely, Z_t can be defined for $t > 0$ by

$$\langle Z_t, g \rangle = \sum_{i \in I} \int_0^{\sigma_i} d\ell_s^t(\zeta^i) g(\hat{W}_s^i) \,,$$

where $\ell_s^t(\zeta^i)$ denotes the local time at level t and at time s of $(\zeta_r^i, r \geq 0)$.

Remarks.

(i) The local time $\ell_s^t(\zeta^i)$ can be defined via the usual approximation

$$\ell_s^t(\zeta^i) = \lim_{\varepsilon \to 0} \frac{1}{\varepsilon} \int_0^s dr \, 1_{(t,t+\varepsilon)}(\zeta_r^i),$$

and $(\ell_s^t(\zeta^i), s \geq 0)$ is a continuous increasing function, for every $i \in I$, a.s.

(ii) The superprocess Z has branching mechanism $2\,u^2$, but a trivial modification will give a superprocess with branching mechanism $\beta\,u^2$, for any choice of $\beta > 0$. Simply observe that, for every $\lambda > 0$, the process $(\lambda Z_t, t \geq 0)$ is a $(\xi, 2\lambda u^2)$-superprocess.

Proof. Let $\mathcal{L}$ denote the random measure on $\mathbb{R}_+ \times E$ defined by

$$\int \mathcal{L}(dt\,dy)h(t)g(y) = \sum_{i \in I} \int_0^{\sigma_i} h(\zeta_s^i)g(\hat{W}_s^i)ds \,,$$

for $h \in \mathcal{B}_{b+}(\mathbb{R}_+)$ and $g \in \mathcal{B}_{b+}(E)$. Suppose that h is compactly supported. By the exponential formula for Poisson measures and then Proposition 3, we get

$$E\left(\exp - \int \mathcal{L}(dt\,dy)h(t)g(y)\right)$$

$$= \exp\left(-\int \mu(dx)\mathbb{N}_x\left(1 - \exp - \int_0^\sigma ds\, h(\zeta_s)g(\hat{W}_s)\right)\right)$$

$$= \exp(-\langle \mu, u_0 \rangle)$$

where $(u_t(x), t \geq 0, x \in E)$ is the unique nonnegative solution of

$$u_t(x) + 2\Pi_{t,x}\left(\int_t^\infty dr\,(u_r(\xi_r))^2\right) = \Pi_{t,x}\left(\int_t^\infty dr\, h(r)g(\xi_r)\right).$$

By comparing with Corollary II.9, we see that the random measure $\mathcal{L}$ has the same distribution as

$$dt\, Z_t'(dy)$$

where Z' is a $(\xi, 2u^2)$-superprocess with $Z_0' = \mu$.

Since Z' is continuous in probability (Proposition II.8) we easily obtain that, for every $t \geq 0$,

$$Z'_t = \lim_{\varepsilon \downarrow 0} \frac{1}{\varepsilon} \int_t^{t+\varepsilon} Z'_r dr \ ,$$

in probability. It follows that for every $t \geq 0$ the limit

$$Z_t(dy) := \lim_{\varepsilon \downarrow 0} \frac{1}{\varepsilon} \int_t^{t+\varepsilon} \mathcal{L}(dr \, dy)$$

exists in probability. Clearly the process Z has the same distribution as Z', and is thus also a $(\xi, 2u^2)$-superprocess started at μ.

Then, if $t > 0$ and $g \in C_{b+}(E)$,

$$\langle Z_t, g \rangle = \lim_{\varepsilon \downarrow 0} \frac{1}{\varepsilon} \int \mathcal{L}(dr \, dy) 1_{[t,t+\varepsilon]}(r) g(y)$$

$$= \lim_{\varepsilon \downarrow 0} \frac{1}{\varepsilon} \sum_{i \in I} \int_0^{\sigma_i} ds \, 1_{[t,t+\varepsilon]}(\zeta_s^i) g(\hat{W}_s^i) \ . \tag{5}$$

Note that there is only a finite number of nonzero terms in the sum over $i \in I$ (for $t > 0$, $\mathbb{N}_x(\sup \zeta_s \geq t) = n(\sup e(s) \geq t) < \infty$). Furthermore, we claim that

$$\lim_{\varepsilon \downarrow 0} \frac{1}{\varepsilon} \int_0^{\sigma} ds \, 1_{[t,t+\varepsilon]}(\zeta_s) g(\hat{W}_s) = \int_0^{\sigma} d\ell_s^t(\zeta) g(\hat{W}_s)$$

in $\mathbb{N}_x$-measure, for every $x \in E$. To verify the claim, note that

$$\int_0^{\sigma} d\ell_s^t(\zeta) g(\hat{W}_s) = \int_0^{\infty} dr \, 1_{\{\tau_r < \infty\}} g(\hat{W}_{\tau_r}),$$

$$\frac{1}{\varepsilon} \int_0^{\sigma} ds \, 1_{[t,t+\varepsilon]}(\zeta_s) g(\hat{W}_s) = \int_0^{\infty} dr \, 1_{\{\tau_r^\varepsilon < \infty\}} g(\hat{W}_{\tau_r^\varepsilon}),$$

where

$$\tau_r = \inf\{s, \ell_s^t > r\}, \qquad \tau_r^\varepsilon = \inf\{s, \frac{1}{\varepsilon} \int_0^s du \, 1_{(t,t+\varepsilon)}(\zeta_u) > r\}.$$

We know that $\tau_r^\varepsilon \longrightarrow \tau_r$ as $\varepsilon \to 0$, $\mathbb{N}_x$ a.e. on $\{\tau_r < \infty\}$. From the continuity properties of W (more specifically, from Lemma 1 (ii)), we get for every $r > 0$

$$\lim_{\varepsilon \to 0} \mathbb{N}_x \left(|g(\hat{W}_{\tau_r^\varepsilon}) 1_{\{\tau_r^\varepsilon < \infty\}} - g(\hat{W}_{\tau_r}) 1_{\{\tau_r < \infty\}}| \right) = 0.$$

The claim follows, and the formula for Z_t in the theorem is then a consequence of (5).

Finally, the first formula in the statement of Theorem 4 is a consequence of the formula for Z_t and the occupation time density formula for Brownian local times.   □

Let us comment on the representation provided by Theorem 4. Define under N_x a measure-valued process $(\mathcal{Z}_t, t > 0)$ by the formula

$$\langle \mathcal{Z}_t, g \rangle = \int_0^\sigma d\ell_s^t(\zeta) g(\hat{W}_s) \,. \tag{6}$$

The "law" of $(\mathcal{Z}_t, t > 0)$ under N_x is sometimes called the canonical measure (of the $(\xi, 2u^2)$-superprocess) with initial point x. Intuitively the canonical measure represents the contributions to the superprocess of the descendants of one single "individual" alive at time 0 at the point x. (This intuitive explanation could be made rigorous by an approximation by discrete branching particle systems in the spirit of Chapter II.) The representation of the theorem can be written as

$$Z_t = \sum_{i \in I} Z_t^i$$

and means (informally) that the population at time t is obtained by superimposing the contributions of the different individuals alive at time 0.

The canonical representation can be derived independently of the Brownian snake approach: Up to some point, it is a special case of the Lévy-Khintchine decomposition for infinitely divisible random measures (see e.g. [Ka]). The advantage of the Brownian snake approach is that it gives the explicit formula (6) for the canonical measure.

Another nice feature of this approach is the fact that it gives simultaneously the associated historical superprocess. Recall from Chapter II that this is the $(\tilde{\xi}, 2u^2)$-superprocess, where $\tilde{\xi}_t = (\xi_r, 0 \leq r \leq t)$ can be viewed as a Markov process with values in $\mathcal{W}$. In fact, with the notation of Theorem 4, the formula

$$\langle \tilde{Z}_t, G \rangle = \sum_{i \in I} \int_0^{\sigma_i} d\ell_s^t(\zeta^i) G(W_s^i)$$

defines a historical superprocess started at μ. The proof of this fact is immediate from Theorem 4 if one observes that the ξ-Brownian snake and the $\tilde{\xi}$-Brownian snake are related in a trivial way.

4   The case of continuous spatial motion

When the spatial motion ξ has (Hölder) continuous sample paths , the Brownian snake has also stronger continuity properties, and the Brownian snake representation of superprocesses easily leads to certain interesting sample path properties.

Recall that $\delta(x, y)$ denotes the metric on E. In this section and the next one, we will assume the following hypothesis, which is stronger than (1).

Assumption (C). *There exist three constants C, $p > 2$, $\varepsilon > 0$ such that, for every $x \in E$ and for every $t \geq 0$,*

$$\Pi_x \left(\sup_{0 \leq r \leq t} \delta(x, \xi_r)^p \right) \leq C \, t^{2+\varepsilon} \ .$$

By the classical Kolmogorov lemma, this implies that the process ξ has (Hölder) continuous sample paths. Note that assumption (C) holds when ξ is Brownian motion or a nice diffusion process in $\mathbb{R}^d$ or on a manifold.

It is then clear that we can construct the ξ-Brownian snake as a process with values in the space of (finite) continuous paths, rather than càdlàg paths as in Section 1. With a slight abuse of notation, we now write $\mathcal{W}$ for the space of all E-valued finite continuous paths and d for the metric on $\mathcal{W}$ defined by

$$d(w, w') = |\zeta_w - \zeta_{w'}| + \sup_{t \geq 0} \delta\big(w(t \wedge \zeta_w), w'(t \wedge \zeta_{w'})\big) \ .$$

Proposition 5. *The process $(W_s, s \geq 0)$ has a continuous modification under $\mathbb{N}_x$ or under $\mathbb{P}_w$ for every $x \in E$, $w \in \mathcal{W}$.*

Remark. We should say more accurately that the measurable modification constructed in Section 1 has continous sample paths, $\mathbb{P}_w$ a.s. or $\mathbb{N}_x$ a.e.

Proof. Recall that paths of (reflected) linear Brownian motion are Hölder continuous with exponent $1/2 - \eta$ for every $\eta > 0$. Fix a function $f \in C(\mathbb{R}_+, \mathbb{R}_+)$ such that for every $T > 0$ and every $\eta \in (0, 1/2)$, there exists a constant $C_{\eta, T}$ with

$$|f(s) - f(s')| \leq C_{\eta, T} \, |s - s'|^{1/2 - \eta} \ , \quad \forall s, s' \in [0, T] \ .$$

Proposition 5 will follow if we can prove that the process $(W_s, s \geq 0)$ has a continuous modification under Θ_w^f, for any w such that $\zeta_w = f(0)$.

Suppose first that $f(0) = 0$, and so $w = x \in E$. By the construction of Section 1, the joint distribution of $(W_s, W_{s'})$ under Θ_x^f is

$$\Pi_x^{f(s)}(dw) R_{m(s, s'), f(s')}(w, dw').$$

Then, for every $s, s' \in [0, T]$, $s \leq s'$,

$$\Theta_x^f \big(d(W_s, W_{s'})^p\big)$$

$$\leq c_p \Big(|f(s) - f(s')|^p + 2 \Pi_x \Big(\Pi_{\xi_{m(s,s')}} \Big(\sup_{0 \leq t \leq (f(s) \vee f(s')) - m(s,s')} \delta(\xi_0, \xi_t)^p \Big) \Big) \Big)$$

$$\leq c_p \Big(|f(s) - f(s')|^p + 2C \, |(f(s) \vee f(s')) - m(s, s')|^{2+\varepsilon} \Big)$$

$$\leq c_p \Big(C_{\eta,T}^p \, |s - s'|^{p(\frac{1}{2}-\eta)} + 2C \, C_{\eta,T}^{2+\varepsilon} \, |s - s'|^{(2+\varepsilon)(\frac{1}{2}-\eta)} \Big),$$

where we used assumption (C) in the second inequality. We can choose $\eta > 0$ small enough so that $p(\frac{1}{2} - \eta) > 1$ and $(2 + \varepsilon)(\frac{1}{2} - \eta) > 1$. The desired result then follows from the classical Kolmogorov lemma.

When $f(0) > 0$, the same argument gives the existence of a continuous modification on every interval $[a, b]$ such that $f(s) > m(0, s)$ for every $s \in (a, b)$. More precisely, the proof of the Kolmogorov lemma shows that this continuous modification satisfies a Hölder condition independent of $[a, b]$ provided that $b \leq K$. On the other hand, if s is such that $f(s) = m(0, s)$ the construction of the Brownian snake shows that $W_s(t) = w(t)$, $\forall t \leq f(s)$, Θ_w^f a.s. Replacing W by a modification, we may assume that the latter property holds simultaneously for all s such that $f(s) = m(0, s)$, Θ_w^f a.s. Then, if $s_1 < s_2$ are not in the same excursion interval of $f(s) - m(0, s)$ away from 0, we simply bound

$$d(W_{s_1}, W_{s_2}) \leq d(W_{s_1}, W_{b_1}) + d(W_{b_1}, W_{a_2}) + d(W_{a_2}, W_{s_2}),$$

where $b_1 = \inf\{r \geq s_1, f(r) = m(0, r)\}$, $a_2 = \sup\{r \leq s_2, f(r) = m(0, r)\}$. The desired result follows easily. □

From now on, we consider only the continuous modification of the process W provided by Proposition 5. As a consequence of the sample path continuity, we obtain that $\mathbb{P}_w$ a.s. (or $\mathbb{N}_x$ a.e.) for every $s < s'$ we have

$$W_s(t) = W_{s'}(t), \qquad \text{for every } t \leq \inf_{s \leq r \leq s'} \zeta_r.$$

For a fixed choice of s and s', this is immediate from the construction of the Brownian snake. The fact that this property holds simultaneously for all $s < s'$ then follows by continuity. We will sometimes refer to the previous property as the *snake property*.

We now state the strong Markov property of W, which is very useful in applications. We denote by $\mathcal{F}_s$ the σ-field generated by W_r, $0 \leq r \leq s$ and as usual we take

$$\mathcal{F}_{s+} = \bigcap_{r > s} \mathcal{F}_r.$$

Theorem 6. *The process $(W_s, \mathbb{P}_w)$ is strong Markov with respect to the filtration $(\mathcal{F}_{s+})$.*

Proof. Let T be a stopping time of the filtration $(\mathcal{F}_{s+})$ such that $T \leq K$ for some $K < \infty$. Let F be bounded and $\mathcal{F}_{T+}$ measurable, and let Ψ be a bounded measurable function on $\mathcal{W}$. It is enough to prove that for every $s > 0$,

$$\mathbb{E}_w\big(F\,\Psi(W_{T+s})\big) = \mathbb{E}_w\big(F\,\mathbb{E}_{W_T}(\Psi(W_s))\big).$$

We may assume that Ψ is continuous. Then,

$$\mathbb{E}_w\big(F\,\Psi(W_{T+s})\big) = \lim_{n\to\infty} \sum_{k=0}^{\infty} \mathbb{E}_w\big(1_{\{\frac{k}{n}\leq T<\frac{k+1}{n}\}} F\,\Psi(W_{\frac{k+1}{n}+s})\big)$$

$$= \lim_{n\to\infty} \sum_{k=0}^{\infty} \mathbb{E}_w\big(1_{\{\frac{k}{n}\leq T<\frac{k+1}{n}\}} F\,\mathbb{E}_{W_{\frac{k+1}{n}}}(\Psi(W_s))\big).$$

In the first equality, we used the continuity of paths and in the second one the ordinary Markov property, together with the fact that $1_{\{k/n\leq T<(k+1)/n\}}\,F$ is $\mathcal{F}_{(k+1)/n}$-measurable. At this point, we need an extra argument. We claim that

$$\lim_{\varepsilon\downarrow 0}\Big(\sup_{t\leq K, t\leq r\leq t+\varepsilon}\big|\mathbb{E}_{W_r}(\Psi(W_s)) - \mathbb{E}_{W_t}(\Psi(W_s))\big|\Big) = 0, \qquad \mathbb{P}_w \text{ a.s.} \qquad (7)$$

Clearly, the desired result follows from (7), because on the set $\{k/n \leq T < (k+1)/n\}$ we can bound

$$\big|\mathbb{E}_{W_{\frac{k+1}{n}}}(\Psi(W_s)) - \mathbb{E}_{W_T}(\Psi(W_s))\big| \leq \sup_{t\leq K, t\leq r\leq t+\frac{1}{n}}\big|\mathbb{E}_{W_r}(\Psi(W_s)) - \mathbb{E}_{W_t}(\Psi(W_s))\big|.$$

To prove (7), we write down explicitly

$$\mathbb{E}_{W_r}(\Psi(W_s)) = \int \gamma_s^{\zeta_r}(da\,db)\int R_{a,b}(W_r, dw')\,\Psi(w'),$$

and a similar expression holds for $\mathbb{E}_{W_t}(\Psi(W_s))$. Set

$$c(\varepsilon) = \sup_{t\leq K, t\leq r\leq t+\varepsilon}|\zeta_r - \zeta_t|$$

and note that $c(\varepsilon)$ tends to 0 as $\varepsilon \to 0$, $\mathbb{P}_w$ a.s. Then observe that if $t \leq K$ and $t \leq r \leq r+\varepsilon$, the paths W_r and W_t coincide at least up to time $(\zeta_t - c(\varepsilon))_+$. Therefore we have

$$R_{a,b}(W_r, dw') = R_{a,b}(W_t, dw')$$

for every $a \leq (\zeta_t - c(\varepsilon))_+$ and $b \geq a$. The claim (7) follows from this observation and the known explicit form of $\gamma_s^{\zeta_r}(da\,db)$. $\qquad\square$

Remark. The strong Markov property holds for W even if the underlying spatial motion ξ is not strong Markov.

Under the assumptions of this section, we now know that the process W is a continuous strong Markov process. Furthermore, every point $x \in E$ is regular for W, in the sense that $\mathbb{P}_x(T_{\{x\}} = 0) = 1$ if $T_{\{x\}} = \inf\{s > 0, W_s = x\}$. (This is trivial from the analogous property for reflected linear Brownian motion.) Thus it makes sense to consider the excursion measure of W away from x, and this excursion measure is immediately identified with $\mathbb{N}_x$.

It is then standard (see e.g. [Bl], Theorem 3.28) that the strong Markov property holds under the excursion measure $\mathbb{N}_x$ in the following form. Let S be a stopping time of the filtration $(\mathcal{F}_{s+})$ such that $S > 0$ $\mathbb{N}_x$ a.e., let G be a nonnegative $\mathcal{F}_{S+}$-measurable variable and let H be a nonnegative measurable function on $C(\mathbb{R}_+, \mathcal{W}_x)$. Then

$$\mathbb{N}_x\big(G\,H(W_{S+s}, s \geq 0)\big) = \mathbb{N}_x\Big(G\,\mathbb{E}_{W_S}\big(H(W_{s \wedge T_{\{x\}}}, s \geq 0)\big)\Big).$$

5   Some sample path properties

In this section, we use the Brownian snake construction to derive certain sample path properties of superprocesses. It is more convenient to consider first the excursion measures $\mathbb{N}_x$. Recall the definition under $\mathbb{N}_x$ of the random measure $\mathcal{Z}_t$

$$\langle \mathcal{Z}_t, g \rangle = \int_0^\sigma d\ell_s^t(\zeta) g(\hat{W}_s) \,.$$

We let supp $\mathcal{Z}_t$ denote the topological support of $\mathcal{Z}_t$ and define the range $\mathcal{R}$ by

$$\mathcal{R} = \overline{\bigcup_{t \geq 0} \text{supp } \mathcal{Z}_t} \,.$$

Theorem 7. *The following properties hold $\mathbb{N}_x$ a.e. for every $x \in E$:*

(i) *The process $(\mathcal{Z}_t, t \geq 0)$ has continuous sample paths.*

(ii) *For every $t > 0$, the set supp $\mathcal{Z}_t$ is a compact subset of E. If ξ is Brownian motion in $\mathbb{R}^d$,*

$$\dim(\text{supp } \mathcal{Z}_t) = 2 \wedge d$$

a.e. on $\{\mathcal{Z}_t \neq 0\}$.

(iii) *The set $\mathcal{R}$ is a connected compact subset of E. If ξ is Brownian motion in $\mathbb{R}^d$,*

$$\dim(\mathcal{R}) = 4 \wedge d \,.$$

Proof. (i) By the joint continuity of Brownian local times, the mapping $t \to d\ell_s^t(\zeta)$ is continuous from $\mathbb{R}_+$ into $M_f(\mathbb{R}_+)$, $\mathbb{N}_x$ a.e. By Proposition 5, $s \to \hat{W}_s$ is also continuous, $\mathbb{N}_x$ a.e. The desired result follows at once.

(ii) For every $t > 0$, supp $\mathcal{Z}_t$ is contained in the set $\{\hat{W}_s, 0 \leq s \leq \sigma\}$ which is compact, again by Proposition 5. Suppose then that ξ is Brownian motion in $\mathbb{R}^d$. Note that, from the definition of $\mathcal{Z}_t$, and the fact that $s \to \ell_s^t$ increases only when $\zeta_s = t$, we have $\mathbb{N}_x$ a.e. for every $t > 0$,

$$\text{supp}(\mathcal{Z}_t) \subset \{\hat{W}_s; s \in [0, \sigma], \zeta_s = t\} \ .$$

It is well known that $\dim\{s \in [0, \sigma], \zeta_s = t\} \leq 1/2$ (the level sets of a linear Brownian motion have dimension $1/2$). Then observe that assumption (C) is satisfied for any integer $p > 4$ with $\varepsilon = \frac{p}{2} - 2$, so that the proof of Proposition 5 yields the bound

$$\Theta_x^f\big(d(W_s, W_{s'})^p\big) \leq C_\eta(f) |s - s'|^{\frac{p}{2}(\frac{1}{2} - \eta)}$$

for $s, s' \in [0, \sigma(f)]$, $n(df)$ a.e. From the classical Kolmogorov lemma, we get that $s \to W_s$ is Hölder continuous with exponent $\frac{1}{4} - \gamma$ for any $\gamma > 0$. Obviously the same holds for the mapping $s \to \hat{W}_s$ and we conclude from well-known properties of Hausdorff dimension that

$$\dim\{\hat{W}_s; s \in [0, \sigma], \zeta_s = t\} \leq 4 \dim\{s \in [0, \sigma], \zeta_s = t\} \leq 2 \ .$$

This gives the upper bound $\dim(\text{supp } \mathcal{Z}_t) \leq 2$. The corresponding lower bound for $d \geq 2$ was derived in an exercise of Chapter II (use also the relation between $\mathcal{Z}_t$ and the law of a $(\xi, 2u^2)$ superprocess at time t). Finally, the case $d = 1$ derives from the case $d \geq 2$ by an easy projection argument: If for $d = 1$ one had $\dim(\text{supp } \mathcal{Z}_t) < 1$ with positive $\mathbb{N}_x$ measure on $\{\mathcal{Z}_t \neq 0\}$, this would contradict the fact that, for $d = 2$, $\dim(\text{supp } \mathcal{Z}_t) = 2$ a.e. on $\{\mathcal{Z}_t \neq 0\}$.

(iii) We first verify that

$$\mathcal{R} = \{\hat{W}_s; s \in [0, \sigma]\} \ , \quad \mathbb{N}_x \ \text{a.e.} \tag{8}$$

The inclusion $\subset$ immediately follows from the fact that

$$\text{supp } \mathcal{Z}_t \subset \{\hat{W}_s; s \in [0, \sigma], \zeta_s = t\}$$

for every $t \geq 0$, a.e. By known properties of local times, the support of the measure $d\ell_s^t(\zeta)$ is exactly $\{s \in [0, \sigma], \zeta_s = t\}$, $\mathbb{N}_x$ a.e., for any fixed $t > 0$. Thus the previous inclusion is indeed an equality for any fixed $t > 0$, $\mathbb{N}_x$ a.e. Hence we have $\mathbb{N}_x$ a.e.

$$\mathcal{R} \supset \{\hat{W}_s; s \in [0, \sigma], \zeta_s \in (0, \infty) \cap \mathbb{Q}\}$$

and the desired result follows since the set in the right side is easily seen to be dense in $\{\hat{W}_s; s \in [0, \sigma]\}$.

From (8) we immediately obtain that $\mathcal{R}$ is compact and connected. Suppose then that ξ is Brownian motion in $\mathbb{R}^d$. The same argument as for (ii) implies that

$$\dim \mathcal{R} \leq 4 \dim[0, \sigma] = 4 .$$

To complete the proof when $d \geq 4$, introduce the *total occupation measure*

$$\langle \mathcal{J}, g \rangle = \int_0^\sigma ds\, g(\hat{W}_s) \left(= \int_0^\infty dt \langle \mathcal{Z}_t, g \rangle \right)$$

which is obviously supported on $\mathcal{R}$. Let $G(x, y) = \gamma_d |y - x|^{2-d}$ be the Green function of Brownian motion in $\mathbb{R}^d$. Using Proposition 2 (ii) and some straightforward calculations, one easily verifies that for every $K > 0$, $\varepsilon > 0$ and $\gamma > 0$,

$$\mathbb{N}_x \left(\iint_{(B(x,K) \backslash B(x,\varepsilon))^2} \frac{\mathcal{J}(dy)\mathcal{J}(dz)}{|y - z|^{4-\gamma}} \right)$$

$$= \mathbb{N}_x \left(\int_0^\sigma \int_0^\sigma \frac{ds\, ds'}{|W_s - W_{s'}|^{4-\gamma}} 1_{B(x,K) \backslash B(x,\varepsilon)}(W_s) 1_{B(x,K) \backslash B(x,\varepsilon)}(W_{s'}) \right)$$

$$= 4 \int_{\mathbb{R}^d} dz\, G(x, z) \int_{(B(x,K) \backslash B(x,\varepsilon))^2} dy\, dy'\, G(z, y) G(z, y') |y - y'|^{\gamma-4},$$

where $B(x, K) = \{y \in \mathbb{R}^d; |y - x| < K\}$. At this point, we need an elementary lemma, whose proof is left to the reader.

Lemma 8. *For every $\delta > 0$, there exists a constant C_δ such that, for every $x, y, y' \in \mathbb{R}^d$ with $\delta < |y - x| < \delta^{-1}$, $\delta < |y' - x| < \delta^{-1}$,*

$$\int_{\mathbb{R}^d} dz\, |z - x|^{2-d} |y - z|^{2-d} |y' - z|^{2-d} \leq \begin{cases} C_\delta (1 + \log^+ \frac{1}{|y-y'|}) & \text{if } d = 4 , \\ C_\delta |y - y'|^{4-d} & \text{if } d \geq 5 . \end{cases}$$

By applying Lemma 8 in the previous calculation, we get for every $\varepsilon > 0$, $K > 0$,

$$\mathbb{N}_x \left(\iint_{(B(x,K) \backslash B(x,\varepsilon))^2} \frac{\mathcal{J}(dy)\mathcal{J}(dz)}{|y - z|^{4-\gamma}} \right) < \infty.$$

Frostman's lemma then implies that

$$\dim \mathcal{R} \geq \dim(\operatorname{supp} \mathcal{J}) \geq 4 .$$

Finally the case $d \leq 3$ is handled again by a projection argument.   □

We now restate Theorem 7 in terms of superprocesses. For $\mu \in \mathcal{M}_f(E)$ let $(Z_t, t \geq 0)$ denote a $(\xi, \beta u^2)$ superprocess started at μ, where $\beta > 0$.

Corollary 9.

(i) *The process $(Z_t, t \geq 0)$ has a continuous modification. (From now on we only deal with this modification.)*

(ii) *A.s. for every $t > 0$, supp Z_t is a compact subset of E. If ξ is Brownian motion in $\mathbb{R}^d$,*

$$\dim(\operatorname{supp} Z_t) = 2 \wedge d \quad a.s. \ on \quad \{Z_t \neq 0\} .$$

(iii) *Let*

$$\mathcal{R}^Z = \bigcup_{\varepsilon > 0} \left(\overline{\bigcup_{t \geq \varepsilon} \operatorname{supp} Z_t} \right) .$$

Then, if ξ is Brownian motion in $\mathbb{R}^d$

$$\dim \mathcal{R}^Z = 4 \wedge d \quad a.s.$$

Remark. The reason for the somewhat strange definition of $\mathcal{R}^Z$ is that one does not want supp μ to be automatically contained in $\mathcal{R}^Z$.

Proof. By a remark following Theorem 4, we may take $\beta = 2$. Then, most assertions follow from Theorem 4 and Theorem 7: Use the representation provided by Theorem 4 and notice that, for every fixed $t > 0$, there are only a finite number of indices $i \in I$ such that $Z_t^i > 0$ (or equivalently sup $\zeta_s^i > t$). There is however a delicate point in the proof of (i). Theorem 7 (i) gives the existence of a continuous modification of $(Z_t, t > 0)$, but the right-continuity at $t = 0$ is not immediate. We may argue as follows. Let g be a bounded nonnegative Lipschitz function on E, and let $v_t(x)$ be the (nonnegative) solution of the integral equation

$$v_t(x) + 2\Pi_x \left(\int_0^t v_{t-s}(\xi_s)^2 ds \right) = \Pi_x \left(g(\xi_t) \right) . \tag{9}$$

Then for every fixed $t > 0$,

$$\exp -\langle Z_r, v_{t-r} \rangle = E \left(\exp -\langle Z_t, g \rangle | Z_r \right)$$

is a martingale intexed by $r \in [0, t]$. By standard results on martingales,

$$\lim_{r\downarrow 0}\langle Z_r, v_{t-r}\rangle$$

exists a.s., at least along rationals. On the other hand, it easily follows from (9) and assumption (C) that $v_t(x)$ converges to $g(x)$ as $t \downarrow 0$, uniformly in $x \in E$. Hence,

$$\limsup_{r\downarrow 0,\, r\in\mathbb{Q}}\langle Z_r, g\rangle \leq \limsup_{r\downarrow 0,\, r\in\mathbb{Q}}\langle Z_r, v_{t-r}\rangle + \varepsilon(t)$$

with $\varepsilon(t) \to 0$ as $t \downarrow 0$, and similarly for the lim inf. We conclude that

$$\lim_{r\downarrow 0}\langle Z_r, g\rangle$$

exists a.s. and the limit must be $\langle \mu, g\rangle$ by the continuity in probability. $\qquad\square$

Let us conclude with some remarks. In the proof of Theorem 7, we noticed that

$$\text{supp}\, \mathcal{Z}_t = \{\hat{W}_s; s \in [0, \sigma], \zeta_s = t\}, \qquad N_x \ \text{ a.e.} \qquad (10)$$

for every fixed $t > 0$. There are exceptional values of t for which this equality fails and supp $\mathcal{Z}_t$ is a proper subset of $\{\hat{W}_s; s \in [0, \sigma], \zeta_s = t\}$. These values of t correspond to the local maxima of the function $s \to \zeta_s$: See the exercise below for a typical example.

Identities (8) and (10) have proved extremely useful to get precise information on supp Z_t and $\mathcal{R}^Z$: See [LP] and [L12] for typical applications to the exact Hausdorff measure of the range and the support of super-Brownian motion.

Exercise. (*Extinction point of quadratic superprocesses*) Let Z be as above a $(\xi, 2u^2)$-superprocess with initial value $Z_0 = \mu$. Set

$$T = \inf\{t, Z_t = 0\} = \sup\{t, Z_t \neq 0\}$$

(the second equality follows from the fact that $\langle Z_t, 1\rangle$ is a Feller diffusion, which is absorbed at 0). Show that there exists an E-valued random variable U such that

$$\lim_{t\uparrow T, t<T} \frac{Z_t}{\langle Z_t, 1\rangle} = \delta_U , \quad \text{a.s.}$$

[*Hint*: In the representation of Theorem 4, let $j \in I$ and $S \in [0, \sigma_j]$ be such that $\zeta_S^j = \sup_i \sup_{s\in[0,\sigma_i]} \zeta_s^i$. Then $U = \hat{W}_S^j$ and $T = \zeta_S^j$.]
Observe that (10) fails for $t = T$.

6   Integrated super-Brownian excursion

In this last section, which will not be used in the following chapters, we discuss the random measure known as integrated super-Brownian excursion (ISE). The motivation for studying this random measure comes from limit theorems showing that ISE arises in the asymptotic behavior of certain models of statistical mechanics (cf Section I.6).

We suppose that the spatial motion ξ is Brownian motion in $\mathbb{R}^d$. Recall from Section 5 the notation $\mathcal{J}$ for the total occupation measure of the Brownian snake under $\mathbb{N}_x$:

$$\langle \mathcal{J}, g \rangle = \int_0^\sigma ds\, g(\hat{W}_s)\,, \qquad g \in \mathcal{B}_+(\mathbb{R}^d).$$

Informally, ISE is $\mathcal{J}$ under $\mathbb{N}_0(\cdot \mid \sigma = 1)$.

To give a cleaner definition, recall the notation $n_{(1)}$ for the law of the normalized Brownian excursion (cf Section III.5). With the notation of Section 1, define a probability measure $\mathbb{N}_x^{(1)}$ on $C(\mathbb{R}_+, \mathbb{R}_+) \times \mathcal{W}^{\mathbb{R}_+}$ by setting

$$\mathbb{N}_x^{(1)}(df\, d\omega) = n_{(1)}(df)\, \Theta_x^f(d\omega).$$

The argument of the proof of Proposition 5 shows that $(W_s, 0 \leq s \leq 1)$ has a continuous modification under $\mathbb{N}_0^{(1)}$.

Definition. *ISE is the random measure $\mathcal{J}$ on $\mathbb{R}^d$ defined under $\mathbb{N}_0^{(1)}$ by*

$$\langle \mathcal{J}, g \rangle = \int_0^1 ds\, g(\hat{W}_s)\,, \qquad g \in \mathcal{B}_+(\mathbb{R}^d).$$

From Theorem 7 (iii) and a scaling argument, it is straightforward to verify that $\dim(\mathrm{supp}\,\mathcal{J}) = 4 \wedge d$ a.s.

Analogously to Proposition 2, one can use Theorem III.6 to get an explicit formula for the moments of ISE. These moment formulas are important in the proof of the limit theorems involving ISE: See Derbez and Slade [DS].

Before stating the result, recall the notation Π_x^θ introduced in Section 2 above. We use the tree formalism described in Section III.2. In particular, a tree T is defined as the set of its vertices, which are elements of $\cup_{n=0}^\infty \{1,2\}^n$. We denote by L_T the set of all leaves of T and if $v \in T$, $v \neq \phi$, we denote by $\tilde{v}$ the father of v.

Proposition 10. *Let $p \geq 1$ be an integer and let $F \in \mathcal{B}_+(\mathcal{W}^p)$. Then,*

$$\mathbb{N}_0^{(1)}\left(\int_{0 \leq s_1 \leq s_2 \leq \cdots \leq s_p \leq 1} ds_1 \ldots ds_p\, F(W_{s_1}, \ldots, W_{s_p}) \right)$$

$$= 2^{p+1} \int \Lambda_p(d\theta)\, L(\theta)\, \exp(-2L(\theta)^2)\, \Pi_0^\theta\big(F(w_1, \ldots, w_p)\big). \tag{11}$$

Let $g \in \mathcal{B}_+(\mathbb{R}^d)$. Then,

$$N_0^{(1)}(\langle \mathcal{J}, g \rangle^p) = p!\, 2^{p+1} \sum_{T \in \mathbb{T}_p} \int_{(\mathbb{R}_+)^T} \prod_{v \in T} dh_v \left(\sum_{v \in T} h_v \right) \exp\left(-2 \left(\sum_{v \in T} h_v \right)^2 \right)$$

$$\times \int_{(\mathbb{R}^d)^T} \prod_{v \in T} dy_v \left(\prod_{v \in T} p_{h_v}(y_{\tilde{v}}, y_v) \right) \prod_{v \in L_T} g(y_v),$$

$$\tag{12}$$

where $y_{\tilde{v}} = 0$ if $v = \phi$ by convention, and $p_t(y, y')$ denotes the Brownian transition density.

Proof. Formula (11) is an immediate consequence of Theorem III.6 and Proposition 2 (i), along the lines of the proof of Proposition 2 (ii). Formula (12) follows as a special case (taking $F(w_1, \ldots, w_p) = g(\hat{w}_1) \ldots g(\hat{w}_p)$) using the construction of Π_0^θ and the definition of $\Lambda_p(d\theta)$.   □

Formula (11) obviously contains more information than (12). For instance, it yields as easily the moment functionals for space-time ISE, which is the random measure $\mathcal{J}^*$ on $\mathbb{R}_+ \times \mathbb{R}^d$ defined under $N_0^{(1)}$ as

$$\langle \mathcal{J}^*, g \rangle = \int_0^1 ds\, g(\zeta_s, \hat{W}_s), \qquad g \in \mathcal{B}_+(\mathbb{R}_+ \times \mathbb{R}^d).$$

The analogue of (12) when $\mathcal{J}(dx_1) \ldots \mathcal{J}(dx_n)$ is replaced by $\mathcal{J}^*(dt_1 dx_1) \ldots \mathcal{J}^*(dt_n dx_n)$ and $g(x_1, \ldots, x_n)$ by $g(t_1, x_1, \ldots, t_n, x_n)$, is obtained by replacing in the right side $g(y_v; v \in L_T)$ by $g(\ell_v, y_v; v \in L_T)$, provided ℓ_v is defined by

$$\ell_v = \sum_{v' \prec v} h_{v'},$$

where $\prec$ denotes the genealogical order on the tree: If $v = (i_1, \ldots, i_n)$, $v' \prec v$ iff $v' = (i_1, \ldots, i_k)$ for some $k \in \{0, 1, \ldots, n\}$.

Remark. The second formula of Proposition 10 can be rewritten in several equivalent ways. Arguing as in the concluding remarks of Chapter III, we may replace the sum over ordered binary trees with p leaves by a sum over (unordered) binary trees with p labelled leaves. The formula is unchanged, except that the factor $p!\, 2^{p+1}$ is replaced by 2^{2p}. In this way, we (almost) get the usual form of the moment functionals of ISE: See Aldous [Al4] or Derbez and Slade [DS]. There are still some extra factors 2 due to the fact that in the usual definition of ISE, $n_{(1)}(df)$ is replaced by its image under the mapping $f \to 2f$. To recover exactly the usual formula, simply replace $p_{h_v}(y_{\tilde{v}}, y_v)$ by $p_{2h_v}(y_{\tilde{v}}, y_v)$.

Chapter V
Exit Measures and the Nonlinear
Dirichlet Problem

In this chapter we use the Brownian snake approach of the previous chapter to construct the exit measure of quadratic superprocesses. In the special case where the spatial motion is Brownian motion in $\mathbb{R}^d$, the exit measure yields a probabilistic solution of the Dirichlet problem associated with the equation $\Delta u = u^2$ in a regular domain. This probabilistic solution plays a major role in further developments that will be presented in the following chapters.

1   The construction of the exit measure

We consider the Brownian snake W of the previous chapter. We assume that the underlying Markov process (ξ_s, Π_x) satisfies the continuity assumption (C) of Section IV.4, so that the process W has continuous sample paths with respect to the metric d.

Let D be an open set in E and fix $x \in D$. For every $w \in \mathcal{W}_x$ set

$$\tau(w) = \inf\{t \in [0, \zeta_w], w(t) \notin D\},$$

where $\inf \emptyset = +\infty$. Define

$$\mathcal{E}^D = \{W_s(\tau(W_s)); s \geq 0, \tau(W_s) < \infty\},$$

so that $\mathcal{E}^D$ is the set of all exit points from D of the paths W_s, for those paths that do exit D. Our goal is to construct $\mathbb{N}_x$ a.e. a random measure that is in some sense uniformly spread over $\mathcal{E}^D$. To avoid trivial cases, we first assume that

$$\Pi_x(\exists t \geq 0, \xi_t \notin D) > 0. \tag{1}$$

We start by constructing a continuous increasing process that increases only on the set $\{s \geq 0, \tau(W_s) = \zeta_s\}$.

Proposition 1. *The formula*

$$L_s^D = \lim_{\varepsilon \downarrow 0} \frac{1}{\varepsilon} \int_0^s dr \, 1_{\{\tau(W_r) < \zeta_r < \tau(W_r) + \varepsilon\}}$$

defines a continuous increasing process $(L_s^D, s \geq 0)$, $\mathbb{N}_x$ *a.e. or* $\mathbb{P}_w$ *a.s. for any* $w \in \mathcal{W}_x$. *The process* $(L_s^D, s \geq 0)$ *is called the exit local time from* D.

Proof. Since $\mathbb{N}_x$ can be viewed as the excursion measure of W away from x, it is enough to prove that the given statement holds under $\mathbb{P}_w$. Indeed, we know from excursion theory that $\mathbb{N}_x(\cdot \mid \sup \zeta_s > h)$ is the law under $\mathbb{P}_x$ of the first excursion of W away from x with "height" greater than h, and so the result under $\mathbb{N}_x$ can be derived from the case of $\mathbb{P}_x$.

We use the following lemma, where $w \in \mathcal{W}_x$ is fixed.

Lemma 2. *Set* $\gamma_s = (\zeta_s - \tau(W_s))^+$ *and* $\sigma_s = \inf\{v \geq 0, \int_0^v dr \, 1_{\{\gamma_r > 0\}} > s\}$. *Then* $\sigma_s < \infty$ *for every* $s \geq 0$, $\mathbb{P}_w$ *a.s., and the process* $\Gamma_s = \gamma_{\sigma_s}$ *is under* $\mathbb{P}_w$ *a reflected Brownian motion started at* $(\zeta_w - \tau(w))^+$.

Proposition 1 easily follows from Lemma 2: Denote by $(\ell_s, s \geq 0)$ the local time at 0 of Γ. Then, $\mathbb{P}_x$ a.s. for every $s \geq 0$,

$$\ell_s = \lim_{\varepsilon \to 0} \frac{1}{\varepsilon} \int_0^s dr \, 1_{\{0 < \Gamma_r < \varepsilon\}}.$$

Set $A_s = \int_0^s dr \, 1_{\{\gamma_r > 0\}}$ and $L_s^D = \ell_{A_s}$. We get

$$L_s^D = \lim_{\varepsilon \downarrow 0} \frac{1}{\varepsilon} \int_0^{A_s} dr \, 1_{\{0 < \Gamma_r < \varepsilon\}} = \lim_{\varepsilon \downarrow 0} \frac{1}{\varepsilon} \int_0^s dr \, 1_{\{0 < \gamma_r < \varepsilon\}} \, ,$$

which is the desired result.   □

Proof of Lemma 2. For every $\varepsilon > 0$, introduce the stopping times

$$S_1^\varepsilon = \inf\{s \geq 0, \zeta_s \geq \tau(W_s) + \varepsilon\} \qquad T_1^\varepsilon = \inf\{s \geq S_1^\varepsilon, \zeta_s \leq \tau(W_s)\}$$

$$S_{n+1}^\varepsilon = \inf\{s \geq T_n^\varepsilon, \zeta_s \geq \tau(W_s) + \varepsilon\} \quad T_{n+1}^\varepsilon = \inf\{s \geq S_{n+1}^\varepsilon, \zeta_s \leq \tau(W_s)\} \, .$$

We first verify that the stopping times S_n^ε and T_n^ε are finite $\mathbb{P}_w$ a.s. By applying the strong Markov property at $\inf\{s \geq 0, \zeta_s = 0\}$, it is enough to consider the case when $w = x$. Still another application of the strong Markov property shows that it is enough to verify that $S_1^\varepsilon < \infty$ a.s. To this end, observe that $\mathbb{P}_x(\zeta_1 \geq \tau(W_1) + \varepsilon) > 0$ (by (1) and because, conditionally on ζ_1, W_1 is a path of ξ with length ζ_1) and apply the strong Markov property at $\inf\{s \geq 1, \zeta_s = 0\}$.

From the snake property and the continuity of $s \to \zeta_s$, one easily gets that the mapping $s \to \gamma_s$ is also continuous. It follows that $\gamma_{S_1^\varepsilon} = \varepsilon \vee (\zeta_w - \tau(w))$ and $\gamma_{S_n^\varepsilon} = \varepsilon$ for $n \geq 2$.

We then claim that, for every $n \geq 1$, we have

$$T_n^\varepsilon = \inf\{s \geq S_n^\varepsilon, \zeta_s = \tau(W_{S_n^\varepsilon})\} .$$

Indeed the snake property implies that for $S_n^\varepsilon \leq r \leq \inf\{s \geq S_n^\varepsilon, \zeta_s = \tau(W_{S_n^\varepsilon})\}$, the paths W_r and $W_{S_n^\varepsilon}$ coincide for $t \leq \tau(W_{S_n^\varepsilon})$, so that $\tau(W_r) = \tau(W_{S_n^\varepsilon})$. This argument also shows that $\gamma_r = \zeta_r - \tau(W_{S_n^\varepsilon})$ for $S_n^\varepsilon \leq r \leq T_n^\varepsilon$.

From the previous observations and the strong Markov property of the Brownian snake, we see that the processes

$$\left(\gamma_{(S_n^\varepsilon + r) \wedge T_n^\varepsilon}, r \geq 0\right)$$

are independent and distributed according to the law of a linear Brownian motion started at ε (at $\varepsilon \vee (\zeta_w - \tau(w))$ for $n = 1$) and stopped when it hits 0. Hence, if

$$\sigma_r^\varepsilon = \inf\left\{s, \int_0^s \sum_{n=1}^\infty 1_{[S_n^\varepsilon, T_n^\varepsilon]}(u)du > r\right\},$$

the process $(\gamma_{\sigma_r^\varepsilon}, r \geq 0)$ is obtained by pasting together a linear Brownian motion started at $\varepsilon \vee (\zeta_w - \tau(w))$ and stopped when it hits 0, with a sequence of independent copies of the same process started at ε. A simple coupling argument shows that $(\gamma_{\sigma_r^\varepsilon}, r \geq 0)$ converges in distribution as $\varepsilon \to 0$ to reflected Brownian motion started at $(\zeta_w - \tau(w))^+$. The lemma follows since it is clear that $\sigma_r^\varepsilon \downarrow \sigma_r$ a.s. for every $r \geq 0$. $\qquad \square$

Definition. *The exit measure $\mathcal{Z}^D$ from D is defined under $\mathbb{N}_x$ by the formula*

$$\langle \mathcal{Z}^D, g \rangle = \int_0^\sigma dL_s^D g(\hat{W}_s) .$$

From Proposition 1 it is easy to obtain that L_s^D increases only on the (closed) set $\{s \in [0, \sigma], \zeta_s = \tau(W_s)\}$. It follows that $\mathcal{Z}^D$ is ($\mathbb{N}_x$ a.e.) supported on $\mathcal{E}^D$.

Let us consider the case when (1) does not hold. Then a first moment calculation using the case $p = 1$ of Proposition IV.2 shows that

$$\int_0^\infty ds \, 1_{\{\tau(W_s) < \infty\}} = 0 , \quad \mathbb{N}_x \text{ a.e.}$$

Therefore the result of Proposition 1 still holds under $\mathbb{N}_x$ with $L_s^D = 0$ for every $s \geq 0$. Consequently, we take $\mathcal{Z}^D = 0$ in that case.

We will need a first moment formula for L^D. With a slight abuse of notation, we also denote by τ the first exit time from D for ξ.

Proposition 3. *Let Π_x^D denote the law of $(\xi_r, 0 \leq r \leq \tau)$ under the subprobability measure $\Pi_x(\cdot \cap \{\tau < \infty\})$. Then, for every $G \in \mathcal{B}_{b+}(\mathcal{W}_x)$,*

$$\mathbb{N}_x\left(\int_0^\sigma dL_s^D G(W_s)\right) = \Pi_x^D(G) \, .$$

In particular, for $g \in \mathcal{B}_{b+}(E)$,

$$\mathbb{N}_x\left(\langle \mathcal{Z}^D, g \rangle\right) = \Pi_x\left(1_{\{\tau < \infty\}} g(\xi_\tau)\right) \, .$$

Proof. We may assume that G is continuous and bounded, and $G(w) = 0$ if $\zeta_w \leq K^{-1}$ or $\zeta_w \geq K$, for some $K > 0$. By Proposition 1,

$$\int_0^\sigma dL_s^D \, G(W_s) = \lim_{\varepsilon \to 0} \frac{1}{\varepsilon} \int_0^\sigma ds \, 1_{\{\tau(W_s) < \zeta_s < \tau(W_s) + \varepsilon\}} \, G(W_s) \qquad (2)$$

$\mathbb{N}_x$ a.e. If we can justify the fact that the convergence (2) also holds in $L^1(\mathbb{N}_x)$, we will get from the case $p = 1$ of Proposition IV.2 (ii):

$$\mathbb{N}_x\left(\int_0^\sigma dL_s^D G(W_s)\right) = \lim_{\varepsilon \to 0} \frac{1}{\varepsilon} \int_0^\infty dh \, \Pi_x\left(1_{\{\tau < h < \tau + \varepsilon\}} G(\xi_r, 0 \leq r \leq h)\right)$$

$$= \Pi_x\left(1_{\{\tau < \infty\}} G(\xi_r, 0 \leq r \leq \tau)\right) \, .$$

It remains to justify the convergence in $L^1(\mathbb{N}_x)$. Because of our assumption on G we may deal with the finite measure $\mathbb{N}_x(\cdot \cap \{\sup \zeta_s > K^{-1}\})$ and so it is enough to prove that

$$\sup_{\varepsilon \in (0,1)} \mathbb{N}_x\left(\left(\frac{1}{\varepsilon} \int_0^\sigma ds \, 1_{\{\tau(W_s) < \zeta_s < \tau(W_s) + \varepsilon\}} G(W_s)\right)^2\right)$$

is finite. This easily follows from the case $p = 2$ of Proposition IV.2 (ii), using now the fact that $G(w) = 0$ if $\zeta_w \geq K$. $\qquad\square$

Remark. Proposition 3 will be considerably extended in Section 4 below.

Let us conclude this section with an important remark. Without any additional effort, the previous construction applies to the more general case of a space-time open set $D \subset \mathbb{R}_+ \times E$, such that $(0, x) \in D$. In this setting, $\mathcal{Z}^D$ is a random measure on $\partial D \subset \mathbb{R}_+ \times E$ such that for $g \in C_{b+}(\partial D)$

$$\langle \mathcal{Z}^D, g \rangle = \lim_{\varepsilon \to 0} \frac{1}{\varepsilon} \int_0^\sigma ds \, 1_{\{\tau(W_s) < \zeta_s < \tau(W_s) + \varepsilon\}} g(\zeta_s, \hat{W}_s)$$

where $\tau(w) = \inf\{t \geq 0, (t, w(t)) \notin D\}$. To see that this more general case is in fact contained in the previous construction, simply replace ξ by the space-time process $\xi'_t = (t, \xi_t)$, which also satisfies assumption (C), and note that the ξ'-Brownian snake is related to the ξ-Brownian snake in a trivial manner. In the special case when $D = D_a = [0, a) \times E$, it is easy to verify that $\mathcal{Z}^{D_a} = \delta_a \otimes \mathcal{Z}_a$, where the measure $\mathcal{Z}_a$ was defined in Section IV.4.

2 The Laplace functional of the exit measure

We will now derive an integral equation for the Laplace functional of the exit measure. This result is the key to the connections with partial differential equations that will be investigated later.

Theorem 4. *For $g \in \mathcal{B}_{b+}(E)$, set*

$$u(x) = \mathbb{N}_x\left(1 - \exp-\langle \mathcal{Z}^D, g\rangle\right), \quad x \in D.$$

The function u solves the integral equation

$$u(x) + 2\Pi_x\left(\int_0^\tau u(\xi_s)^2 ds\right) = \Pi_x\left(1_{\{\tau<\infty\}}g(\xi_\tau)\right). \tag{3}$$

Our proof of Theorem 4 is based on a lemma of independent interest, which has many other applications. Another more computational proof, in the spirit of the proof of Proposition IV.3, would rely on calculations of moments of the exit measure (cf Section 4). Still another method would consist in writing $\langle \mathcal{Z}^D, g\rangle$ as a limit of integrals of the form $\int dt\langle \tilde{\mathcal{Z}}_t, h_\varepsilon\rangle$ (with $(\tilde{\mathcal{Z}}_t, t \geq 0)$ corresponding to the historical superprocess, as in Section IV.3) and then using the form of the Laplace functional of these integrals obtained in Chapter II.

Before stating our key lemma, we need some notation. We fix $w \in \mathcal{W}_x$ with $\zeta_w > 0$ and consider the Brownian snake under $\mathbb{P}_w$. We set

$$T_0 = \inf\{s \geq 0, \zeta_s = 0\}$$

and denote by (α_i, β_i), $i \in I$ the excursion intervals of $\zeta_s - \inf_{[0,s]} \zeta_r$ before time T_0. In other words, (α_i, β_i), $i \in I$ are the connected components of the open set $[0, T_0] \cap \{s \geq 0, \zeta_s > \inf_{[0,s]} \zeta_r\}$. Then, for every $i \in I$ we define $W^i \in C(\mathbb{R}_+, \mathcal{W})$ by setting for every $s \geq 0$,

$$W^i_s(t) = W_{(\alpha_i+s)\wedge\beta_i}(\zeta_{\alpha_i} + t), \qquad 0 \leq t \leq \zeta^i_s := \zeta_{(\alpha_i+s)\wedge\beta_i} - \zeta_{\alpha_i}.$$

From the snake property we have in fact $W^i \in C(\mathbb{R}_+, \mathcal{W}_{w(\zeta_{\alpha_i})})$.

Lemma 5. *The point measure*

$$\sum_{i \in I} \delta_{(\varsigma_{\alpha_i}, W^i)}$$

is under $\mathbb{P}_w$ *a Poisson point measure on* $\mathbb{R}_+ \times C(\mathbb{R}_+, \mathcal{W})$ *with intensity*

$$2\, 1_{[0,\varsigma_w]}(t)dt\, \mathbb{N}_{w(t)}(dw) \,.$$

Proof. A well-known theorem of Lévy states that, if $(\beta_t, t \geq 0)$ is a linear Brownian motion started at a, the process $\beta_t - \inf_{[0,t]} \beta_r$ is a reflected Brownian motion whose local time at 0 is $2(a - \inf_{[0,t]} \beta_r)$. From this and excursion theory, it follows that the point measure

$$\sum_{i \in I} \delta_{(\varsigma_{\alpha_i}, \varsigma^i)}$$

is under $\mathbb{P}_w$ a Poisson point measure with intensity

$$2\, 1_{[0,\varsigma_w]}(t)dt\, n(de) \,.$$

It remains to combine this result with the spatial displacements.

To this end, fix a function $f \in C(\mathbb{R}_+, \mathbb{R}_+)$ such that $f(0) = \varsigma_w$, $T_0(f) := \inf\{t, f(t) = 0\} < \infty$ and f is locally Hölder with exponent $\frac{1}{2} - \gamma$ for every $\gamma > 0$. Recall the notation Θ_w^f from Chapter IV and note that Θ_w^f can be viewed as a probability measure on $C(\mathbb{R}_+, \mathcal{W})$ (see the proof of Proposition IV 5). Denote by $e_j, j \in J$ the excursions of $f(s) - \inf_{[0,s]} f(r)$ away from 0 before time $T_0(f)$, by $(a_j, b_j), j \in J$ the corresponding time intervals, and define for every $j \in J$

$$W_s^j(t) = W_{(a_j+s) \wedge b_j}(f(a_j) + t)\,, \quad 0 \leq t \leq f((a_j + s) \wedge b_j) - f(a_j)\,,$$

From the definition of Θ_w^f, it is easily verified that the processes $W^j, j \in J$ are independent under Θ_w^f, with respective distributions $\Theta_{w(f(a_j))}^{e_j}$.

Let $F \in \mathcal{B}_{b+}(\mathbb{R}_+ \times C(\mathbb{R}_+, \mathcal{W}))$ be such that $F(t, w) = 0$ if $\sup \varsigma_s(w) \leq \gamma$, for some $\gamma > 0$. Recall the notation $P_r(df)$ for the law of reflected Brownian motion started at r. By using the last observation and then the beginning of the proof, we get

$$\mathbb{E}_w\left(\exp - \sum_{i \in I} F(\varsigma_{\alpha_i}, W^i)\right) = \int P_{\varsigma_w}(df)\Theta_w^f\left(\exp - \sum_{j \in J} F(f(a_j), W^j)\right)$$

$$= \int P_{\varsigma_w}(df) \prod_{j \in J} \Theta_{w(f(a_j))}^{e_j}\left(e^{-F(f(a_j), \cdot)}\right)$$

$$= \exp - 2 \int_0^{\varsigma_w} dt \int n(de)\Theta_{w(t)}^e\left(1 - e^{-F(t, \cdot)}\right)$$

$$= \exp - 2 \int_0^{\varsigma_w} dt\, \mathbb{N}_{w(t)}\left(1 - e^{-F(t, \cdot)}\right) \,.$$

The third equality is the exponential formula for Poisson measures, and the last one is the definition of $\mathbb{N}_x$. This completes the proof. $\qquad\square$

Proof of Theorem 4. By the definition of $\mathcal{Z}^D$, we have

$$u(x) = \mathbb{N}_x\left(1 - \exp - \int_0^\sigma dL_s^D g(\hat{W}_s)\right)$$

$$= \mathbb{N}_x\left(\int_0^\sigma dL_s^D g(\hat{W}_s) \exp\left(-\int_s^\sigma dL_r^D g(\hat{W}_r)\right)\right)$$

$$= \mathbb{N}_x\left(\int_0^\sigma dL_s^D g(\hat{W}_s)\mathbb{E}_{W_s}\left(\exp - \int_0^{T_0} dL_r^D g(\hat{W}_r)\right)\right)$$

using the strong Markov property under $\mathbb{N}_x$ in the last equality. Let $w \in \mathcal{W}_x$ be such that $\zeta_w = \tau(w)$. From Lemma 5, we have

$$\mathbb{E}_w\left(\exp - \int_0^{T_0} dL_r^D g(\hat{W}_r)\right)$$

$$= \mathbb{E}_w\left(\exp - \sum_{i\in I}\int_{\alpha_i}^{\beta_i} dL_r^D g(\hat{W}_r)\right)$$

$$= \exp\left(-2\int_0^{\zeta_w} dt\, \mathbb{N}_{w(t)}\left(1 - \exp - \int_0^\sigma dL_r^D g(\hat{W}_r)\right)\right)$$

$$= \exp\left(-2\int_0^{\zeta_w} dt\, u\big(w(t)\big)\right).$$

Hence,

$$u(x) = \mathbb{N}_x\left(\int_0^\sigma dL_s^D g(\hat{W}_s) \exp\left(-2\int_0^{\zeta_s} dt\, u\big(W_s(t)\big)\right)\right)$$

$$= \Pi_x\left(1_{\{\tau<\infty\}} g(\xi_\tau) \exp\left(-2\int_0^\tau dt\, u(\xi_t)\right)\right)$$

by Proposition 3. The proof is now easily completed by the usual Feynman-Kac argument:

$u(x)$

$$= \Pi_x\left(1_{\{\tau<\infty\}} g(\xi_\tau)\right) - \Pi_x\left(1_{\{\tau<\infty\}} g(\xi_\tau)\left(1 - \exp -2\int_0^\tau dt\, u(\xi_t)\right)\right)$$

$$= \Pi_x\left(1_{\{\tau<\infty\}} g(\xi_\tau)\right) - 2\Pi_x\left(1_{\{\tau<\infty\}} g(\xi_\tau)\int_0^\tau dt\, u(\xi_t) \exp\left(-2\int_t^\tau dr\, u(\xi_r)\right)\right)$$

$$= \Pi_x\left(1_{\{\tau<\infty\}} g(\xi_\tau)\right) - 2\Pi_x\left(\int_0^\tau dt\, u(\xi_t)\Pi_{\xi_t}\left(1_{\{\tau<\infty\}} g(\xi_\tau) \exp\left(-2\int_0^\tau dr\, u(\xi_r)\right)\right)\right)$$

$$= \Pi_x\left(1_{\{\tau<\infty\}} g(\xi_\tau)\right) - 2\Pi_x\left(\int_0^\tau dt\, u(\xi_t)^2\right). \qquad\square$$

3   The probabilistic solution of the nonlinear Dirichlet problem

In this section, we assume that ξ is Brownian motion in $\mathbb{R}^d$. The results however could easily be extended to an elliptic diffusion process in $\mathbb{R}^d$ or on a manifold. We say that $y \in \partial D$ is regular for D^c if

$$\Pi_y\big(\inf\{t > 0, \xi_t \notin D\} = 0\big) = 1 .$$

The open set D is called regular if every point $y \in \partial D$ is regular for D^c. We say that a real-valued function u defined on D solves $\Delta u = 4u^2$ in D if u is of class C^2 on D and the equality $\Delta u = 4u^2$ holds pointwise on D.

Theorem 6. *Let D be a domain in $\mathbb{R}^d$ and let $g \in \mathcal{B}_{b+}(\partial D)$. For every $x \in D$, set $u(x) = N_x(1 - \exp -\langle \mathcal{Z}^D, g\rangle)$. Then u solves $\Delta u = 4u^2$ in D. If in addition D is regular and g is continuous, then u solves the problem*

$$\begin{cases} \Delta u = 4u^2 & \text{in } D \\ u_{|\partial D} = g \end{cases} \tag{4}$$

where the notation $u_{|\partial D} = g$ means that for every $y \in \partial D$,

$$\lim_{D \ni x \to y} u(x) = g(y) .$$

Proof. First observe that, by (3),

$$u(x) \le \Pi_x\big(1_{\{\tau < \infty\}} g(\xi_\tau)\big) \le \sup_{y \in \partial D} g(y) ,$$

so that u is bounded in D. Let B be a ball whose closure is contained in D, and denote by τ_B the first exit time from B. From (3) and the strong Markov property at time τ_B we get for $x \in B$

$$u(x) + 2\Pi_x\left(\int_0^{\tau_B} u(\xi_s)^2 ds\right) + 2\Pi_x\left(\Pi_{\xi_{\tau_B}}\left(\int_0^\tau u(\xi_s)^2 ds\right)\right)$$
$$= \Pi_x\big(\Pi_{\xi_{\tau_B}}\big(1_{\{\tau < \infty\}} g(\xi_\tau)\big)\big).$$

By combining this with formula (3) applied with $x = \xi_{\tau_B}$, we arrive at

$$u(x) + 2\Pi_x\left(\int_0^{\tau_B} u(\xi_s)^2 ds\right) = \Pi_x\big(u(\xi_{\tau_B})\big) . \tag{5}$$

The function $h(x) = \Pi_x\big(u(\xi_{\tau_B})\big)$ is harmonic in B, so that h is of class C^2 and $\Delta h = 0$ in B. Set

$$f(x) := \Pi_x\left(\int_0^{\tau_B} u(\xi)^2 ds\right) = \int_B dy\, G_B(x,y) u(y)^2$$

where G_B is the Green function of Brownian motion in B. Since u is measurable and bounded, Theorem 6.6 of [PS] shows that f is continuously differentiable in B, and so is u since $u = h - 2f$. Then again by Theorem 6.6 of [PS], the previous formula for f implies that f is of class C^2 in B and $-\frac{1}{2}\Delta f = u^2$ in B, which leads to the desired equation for u.

For the second part of the theorem, suppose first that D is bounded, and let $y \in \partial D$ be regular for D^c. Then, if g is continuous at y, it is well known that

$$\lim_{D \ni x \to y} \Pi_x\big(g(B_\tau)\big) = g(y) \; .$$

On the other hand, we have also

$$\limsup_{D \ni x \to y} \Pi_x\left(\int_0^\tau u(\xi_s)^2 ds\right) \le \big(\sup_{x \in D} u(x)\big)^2 \limsup_{D \ni x \to y} \Pi_x(\tau) = 0 \; .$$

Thus (3) implies that

$$\lim_{D \ni x \to y} u(x) = g(y) \; .$$

When D is unbounded, a similar argument applies after replacing D by $D \cap B$, where B is now a large ball: Argue as in the derivation of (5) to verify that for $x \in D \cap B$,

$$u(x) + 2\Pi_x\left(\int_0^{\tau_{D \cap B}} u(\xi_s)^2 ds\right) = \Pi_x\big(1_{\{\tau \le \tau_B\}} g(\xi_\tau)\big) + \Pi_x\big(1_{\{\tau_B < \tau\}} u(\xi_{\tau_B})\big)$$

and then follow the same route as in the bounded case.   □

The nonnegative solution of the problem (4) is always unique. When D is bounded, this is a consequence of the following analytic lemma. (In the unbounded case, see the exercise below.)

Lemma 7. (Comparison principle) *Let $h : \mathbb{R}_+ \to \mathbb{R}_+$ be a monotone increasing function. Let D be a bounded domain in $\mathbb{R}^d$ and let u, v be two nonnegative functions of class C^2 on D such that $\Delta u \ge h(u)$ and $\Delta v \le h(v)$. Suppose that for every $y \in \partial D$,*

$$\limsup_{D \ni x \to y} \big(u(x) - v(x)\big) \le 0 \; .$$

Then $u \le v$.

Proof. Set $f = u - v$ and $D' = \{x \in D, f(x) > 0\}$. If D' is not empty, we have

$$\Delta f(x) \ge h\big(u(x)\big) - h\big(v(x)\big) \ge 0$$

for every $x \in D'$. Furthermore, it follows from the assumption and the definition of D' that

$$\limsup_{D' \ni x \to z} f(x) \leq 0$$

for every $z \in \partial D'$. Then the classical maximum principle implies that $f \leq 0$
on D', which is a contradiction. $\qquad\square$

Corollary 8. *Let D be a domain in $\mathbb{R}^d$ and let U be a bounded regular subdomain
of D whose closure is contained in D. Then, if u is a nonnegative solution of
$\Delta u = 4u^2$ in D, we have for every $x \in U$*

$$u(x) = \mathbb{N}_x\big(1 - \exp{-\langle \mathcal{Z}^U, u\rangle}\big).$$

Proof. For every $x \in U$, set

$$v(x) = \mathbb{N}_x\big(1 - \exp{-\langle \mathcal{Z}^U, u\rangle}\big).$$

By Theorem 6, v solves $\Delta v = 4v^2$ in U with boundary value $v_{|\partial U} = u_{|\partial U}$. By
Lemma 7, we must have $v(x) = u(x)$ for every $x \in U$. $\qquad\square$

The last proposition of this section provides some useful properties of nonnegative solutions of $\Delta u = 4u^2$ in a domain. For $x \in \mathbb{R}^d$ and $\varepsilon > 0$, we
denote by $B(x, \varepsilon)$ the open ball of radius ε centered at x. Recall the notation
$\mathcal{R} = \{\hat{W}_s, 0 \leq s \leq \sigma\}$ from Chapter IV.

Proposition 9.
   (i) *There exists a positive constant c_d such that for every $x \in \mathbb{R}^d$ and $\varepsilon > 0$,*

$$\mathbb{N}_x\big(\mathcal{R} \cap B(x, \varepsilon)^c \neq \emptyset\big) = c_d \varepsilon^{-2}.$$

   (ii) *Let u be a nonnegative solution of $\Delta u = 4u^2$ in the domain D. Then for
   every $x \in D$,*

$$u(x) \leq c_d \operatorname{dist}(x, \partial D)^{-2}.$$

   (iii) *The set of all nonnegative solutions of $\Delta u = 4u^2$ in D is closed under
   pointwise convergence.*

Proof. (i) By translation invariance we may assume that $x = 0$. We then use
a scaling argument. For $\lambda > 0$, the law under $n(de)$ of $e_\lambda(s) = \lambda^{-1}e(\lambda^2 s)$ is
$\lambda^{-1}n$. It easily follows that the law under $\mathbb{N}_0$ of $W_s^{(\varepsilon)}(t) = \varepsilon^{-1}W_{\varepsilon^4 s}(\varepsilon^2 t)$ is
$\varepsilon^{-2}\mathbb{N}_0$. Then, with an obvious notation,

$$\mathbb{N}_0\big(\mathcal{R} \cap B(0, \varepsilon)^c \neq \emptyset\big) = \mathbb{N}_0\big(\mathcal{R}^{(\varepsilon)} \cap B(0, 1)^c \neq \emptyset\big)$$

$$= \varepsilon^{-2}\mathbb{N}_0\big(\mathcal{R} \cap B(0, 1)^c \neq \emptyset\big).$$

It remains to verify that $N_0(\mathcal{R} \cap B(0,1)^c \neq \emptyset) < \infty$. If this were not true, excursion theory would imply that P_0 a.s., infinitely many excursions of the Brownian snake exit the ball $B(0,1)$ before time 1. Clearly this would contradict the continuity of $s \to W_s$ under P_0.

(ii) Let $x \in D$ and $r > 0$ be such that $\bar{B}(x,r) \subset D$. By Corollary 8, we have for every $y \in B(x,r)$

$$u(y) = N_y\left(1 - \exp-\langle \mathcal{Z}^{B(x,r)}, u \rangle\right) .$$

In particular,

$$u(x) \leq N_x\left(\mathcal{Z}^{B(x,r)} \neq 0\right) \leq N_x\left(\mathcal{R} \cap B(x,r)^c \neq \emptyset\right) = c_d \, r^{-2} .$$

In the second inequality we used the fact that $\mathcal{Z}^{B(x,r)}$ is supported on $\mathcal{E}^{B(x,r)} \subset \mathcal{R} \cap B(x,r)^c$.

(iii) Let (u_n) be a sequence of nonnegative solutions of $\Delta u = 4\,u^2$ in D such that $u_n(x) \longrightarrow u(x)$ as $n \to \infty$ for every $x \in D$. Let U be an open ball whose closure is contained in D. By Corollary 8, for every $n \geq 1$ and $x \in U$,

$$u_n(x) = N_x\left(1 - \exp-\langle \mathcal{Z}^U, u_n \rangle\right).$$

Note that $N_x(\mathcal{Z}^U \neq 0) < \infty$ (by (i)) and the functions u_n are uniformly bounded on ∂U (by (ii)). Hence we can pass to the limit in the previous formula and get $u(x) = N_x\left(1 - \exp-\langle \mathcal{Z}^U, u \rangle\right)$ for $x \in U$. The desired result then follows from Theorem 6. $\qquad\square$

Let us conclude this section with the following remark. Theorem 4 could be applied as well to treat parabolic problems for the operator $\Delta u - 4u^2$. To this end we need only replace the Brownian motion ξ by the space-time process (t, ξ_t). If we make this replacement and let $D \subset \mathbb{R}_+ \times \mathbb{R}^d$ be a space-time domain, and $g \in \mathcal{B}_{b+}(\partial D)$, the formula

$$u(t,x) = N_{t,x}\left(1 - \exp-\langle \mathcal{Z}^D, g \rangle\right)$$

gives a solution of

$$\frac{\partial u}{\partial t} + \frac{1}{2}\Delta u - 2u^2 = 0$$

in D. Furthermore, u has boundary condition g under suitable conditions on D and g. The proof proceeds from the integral equation (3) as for Theorem 4. In the following chapters we will concentrate on elliptic equations, but most of the results have analogues for parabolic problems.

Exercise. Prove that Theorem 4 remains true if g is unbounded, and even if g takes values in $[0, \infty]$ (the special case $g = +\infty$ will be relevant in the next chapter).

[*Hint:* Let $g_n = g \wedge n$ and observe from Proposition 7 (ii) that the functions $u_n(x) = N_x(1 - \exp - \langle \mathcal{Z}^D, g_n \rangle)$ are uniformly bounded on compact subsets of D. Then pass to the limit $n \to \infty$ in equation (5) written for u_n.]

Exercise. Prove that, provided D is regular and g is continuous, the uniqueness of the nonnegative solution of (4) also holds when D is unbounded, even if g is unbounded (but finite-valued!).

[*Hint:* For every $p \geq 1$ set $D_p = D \cap B(0, p)$ and for $x \in D_p$,

$$u_p(x) = N_x\left(1 - \langle \mathcal{Z}^{D_p}, 1_{\partial D} g \rangle\right),$$

$$v_p(x) = N_x\left(1 - \langle \mathcal{Z}^{D_p}, 1_{\partial D} g + 1_{\partial D_p \setminus \partial D} \cdot \infty \rangle\right).$$

Then, if u solves (4), $u_p \leq u \leq v_p$ in D_p. Furthermore, $v_p - u_p \to 0$ as $p \to \infty$.]

4   Moments of the exit measure

In this section, we consider again a general Markov process ξ satisfying the continuity assumptions of Section IV.4. Our goal is to derive explicit formulas analogous to Proposition IV.2 for the moments of the exit measures. These formulas will be used in the applications developed in Chapter VII.

We start with some notation. We fix an integer $p \geq 1$. We slightly extend the definitions of Section III.4 by allowing the value $+\infty$ for the marks attached to the leaves of a marked tree $\theta \in \mathcal{T}_p$. We then define a measure Λ_p^∞ on the set $\mathcal{T}_p$ by the formula

$$\int \Lambda_p^\infty(d\theta) \, \Phi(\theta) = \sum_{T \in \mathbb{T}_p} \int \prod_{v \in T \setminus L_T} dh_v \prod_{v \in L_T} \delta_\infty(dh_v) \, \Phi(T, \{h_v, v \in T\}),$$

where L_T is the set of leaves of T. In other words, we prescribe the value $+\infty$ for the marks of all leaves, but keep the Lebesgue measure for the other marks. Notice that the construction of Π_x^θ in Section IV.2 still makes sense, now as a probability measure on $C(\mathbb{R}_+, E)^p$, when θ is a marked tree such that the mark of every leaf is $+\infty$.

Let $\theta = (T, \{h_v, v \in T\})$ be a marked tree with p leaves and let $v_1, \ldots, v_p$ be the leaves of T listed in lexicographical order. For every $i \in \{1, \ldots, p\}$, set

$$\alpha_i = \sum_{v \prec v_i, v \neq v_i} h_v,$$

where $\prec$ denotes as previously the genealogical order on the tree. Informally, α_i is the birth height of vertex v_i. Finally, if D is an open set in E and $x \in D$, we let $\Pi_x^{\theta,D}$ be the subprobability measure on $\mathcal{W}_x^p$ defined as the law of

$$((w_1(t), 0 \le t \le \tau(w_1)), \dots, (w_p(t), 0 \le t \le \tau(w_p)))$$

under the measure

$$\prod_{i=1}^{p} 1_{\{\alpha_i \le \tau(w_i) < \infty\}} \, \Pi_x^\theta(dw_1 \dots dw_p).$$

Theorem 10. *For any* $F \in \mathcal{B}_+(\mathcal{W}_x^p)$,

$$\mathbb{N}_x \left(\int_{\{0 \le s_1 \le \dots \le s_p \le \sigma\}} dL_{s_1}^D \dots dL_{s_p}^D \, F(W_{s_1}, \dots, W_{s_p}) \right) = 2^{p-1} \int \Lambda_p^\infty(d\theta) \Pi_x^{\theta,D}(F).$$

In particular, for any $g \in \mathcal{B}_+(E)$,

$$\mathbb{N}_x \left(\langle \mathcal{Z}^D, g \rangle^p \right) = 2^{p-1} p! \int \Lambda_p^\infty(d\theta) \int \Pi_x^{\theta,D}(dw_1 \dots dw_p) \, g(\hat{w}_1) \dots g(\hat{w}_p).$$

Proof. We may assume that F is continuous and bounded and that $F(w_1, \dots, w_p) = 0$ if $\zeta_{w_i} \le \delta$ or $\zeta_{w_i} \ge K$ for some $i \in \{1, \dots, p\}$ (here δ and K are two positive constants). By Proposition 1, we have

$$\int_{\{0 \le s_1 \le \dots \le s_p \le \sigma\}} dL_{s_1}^D \dots dL_{s_p}^D \, F(W_{s_1}, \dots, W_{s_p}) = \lim_{\varepsilon \to 0} A_\varepsilon, \qquad \mathbb{N}_x \text{ a.e.} \quad (6)$$

where

$$A_\varepsilon = \varepsilon^{-p} \int_{\{0 \le s_1 \le \dots \le s_p \le \sigma\}} ds_1 \dots ds_p \, F(W_{s_1}, \dots, W_{s_p}) \prod_{i=1}^{p} 1_{\{\tau(W_{s_i}) < \zeta_{s_i} < \tau(W_{s_i}) + \varepsilon\}}.$$

Using Proposition IV.2 and then the definition of Λ_p^∞, we have

$$2^{-(p-1)} \mathbb{N}_x(A_\varepsilon)$$

$$= \varepsilon^{-p} \int \Lambda_p(d\theta) \int \Pi_x^\theta(dw_1 \dots dw_p) \, F(w_1, \dots, w_p) \prod_{i=1}^{p} 1_{\{\tau(w_i) < \zeta_{w_i} < \tau(w_i) + \varepsilon\}}$$

$$= \varepsilon^{-p} \int \Lambda_p^\infty(d\theta) \int \Pi_x^\theta(dw_1 \dots dw_p) \int_{[0,\infty)^p} dh_1 \dots dh_p$$

$$F(w_1[0, \alpha_1 + h_1], \dots, w_p[0, \alpha_p + h_p]) \prod_{i=1}^{p} 1_{\{\tau(w_i) < \alpha_i + h_i < \tau(w_i) + \varepsilon\}}$$

where $w[0, t]$ stands for the restriction of w to $[0, t]$. For fixed θ and $w_1, \ldots, w_p$, we have

$$\lim_{\varepsilon \to 0} \varepsilon^{-p} \int_{[0,\infty)^p} dh_1 \ldots dh_p \, F(w_1[0, \alpha_1 + h_1], \ldots, w_p[0, \alpha_p + h_p])$$

$$\times \prod_{i=1}^{p} 1_{\{\tau(w_i) < \alpha_i + h_i < \tau(w_i) + \varepsilon\}}$$

$$= F(w_1[0, \tau(w_1)], \ldots, w_p[0, \tau(w_p)]) \prod_{i=1}^{p} 1_{\{\alpha_i \leq \tau(w_i) < \infty\}}.$$

We can then use dominated convergence to pass to the limit $\varepsilon \to 0$ in the previous formula for $N_x(A_\varepsilon)$. Note that the assumptions on F allow us to restrict our attention to the set $\{\alpha_i \leq K, 1 \leq i \leq p\}$, which has finite Λ_p^∞-measure. Recalling the definition of $\Pi_x^{\theta, D}$, we get

$$\lim_{\varepsilon \to 0} N_x(A_\varepsilon) = 2^{p-1} \int \Lambda_p^\infty(d\theta) \Pi_x^{\theta, D}(F). \tag{7}$$

In particular the collection $(A_\varepsilon, \varepsilon \in (0, 1))$ is bounded in $L^1(N_x)$. By replacing p by $2p$ we see similarly that this collection is bounded in $L^2(N_x)$. Hence the convergence (6) holds in $L^1(N_x)$ (note that we can restrict our attention to the set $\{\sup \zeta_s \geq \delta\}$, which has finite N_x-measure). The first formula of the theorem follows from (6) and (7), and the second formula is clearly a special case of the first one. $\qquad \square$

Remark. We could also have derived Theorem 10 from Theorem 4. The previous approach is more appealing to intuition as it explains why the moment formulas involve tree structures.

Exercise. Verify that for $g \in \mathcal{B}_+(E)$, for every $p \geq 2$,

$$N_x(\langle \mathcal{Z}^D, g \rangle^p) = 2 \sum_{j=1}^{p-1} \binom{p}{j} \Pi_x \left(\int_0^\tau dt \, N_{\xi_t}(\langle \mathcal{Z}^D, g \rangle^j) \, N_{\xi_t}(\langle \mathcal{Z}^D, g \rangle^{p-j}) \right)$$

(compare with formula (3) of Chapter IV). Give another proof of Theorem 4 along the lines of the proof of Proposition IV.3.

Chapter VI
Polar Sets and Solutions with Boundary Blow-up

In this chapter, we consider the case when the spatial motion ξ is Brownian motion in $\mathbb{R}^d$ and we continue our investigation of the connections between the Brownian snake and the partial differential equation $\Delta u = 4u^2$. In particular, we show that the maximal nonnegative solution in a domain D can be interpreted as the hitting probability of D^c for the Brownian snake. We then combine analytic and probabilistic techniques to give a characterization of polar sets for the Brownian snake or equivalently for super-Brownian motion. In the last two sections, we investigate two problems concerning solutions with boundary blow-up. We first give a complete characterization of those domains in $\mathbb{R}^d$ in which there exists a (nonnegative) solution which blows up everywhere at the boundary. This analytic result is equivalent to a Wiener test for the Brownian snake or for super-Brownian motion. Finally, in the case of a regular domain, we give sufficient conditions that ensure the uniqueness of the solution with boundary blow-up.

1   Solutions with boundary blow-up

Throughout this chapter, the spatial motion ξ is Brownian motion in $\mathbb{R}^d$. Let us summarize some key results of the previous chapter (Theorem V.4, Theorem V.6, Lemma V.7, Corollary V.8).

(A) If D is a domain in $\mathbb{R}^d$, and $g \in \mathcal{B}_{b+}(\partial D)$, the function $u(x) = \mathbb{N}_x(1 - \exp -\langle \mathcal{Z}^D, g \rangle)$, for $x \in D$, solves the integral equation

$$u(x) + 2\Pi_x \left(\int_0^\tau u(\xi_s)^2 ds \right) = \Pi_x \left(1_{\{\tau < \infty\}} g(\xi_\tau) \right) , \tag{1}$$

(where τ is the first exit time from D) and the differential equation $\Delta u = 4u^2$ in D. If in addition D is regular and g is continuous, u is the unique nonnegative solution of the problem

$$\begin{cases} \Delta u = 4u^2 , & \text{in } D , \\ u_{|\partial D} = g . \end{cases} \qquad (2)$$

(B) If D is a domain in $\mathbb{R}^d$ and U is a bounded regular subdomain of D, whose closure is contained in D, then for any nonnegative solution u of $\Delta u = 4u^2$ in D we have

$$u(x) = \mathbb{N}_x(1 - \exp -\langle \mathcal{Z}^U, u \rangle) , \quad x \in U .$$

Proposition 1. *Let D be a bounded regular domain. Then $u_1(x) = \mathbb{N}_x(\mathcal{Z}^D \neq 0)$, $x \in D$ is the minimal nonnegative solution of the problem*

$$\begin{cases} \Delta u = 4u^2 , & \text{in } D , \\ u_{|\partial D} = +\infty . \end{cases} \qquad (3)$$

Proof. First note that $u_1(x) < \infty$ by Proposition V.9 (i). For every $n \geq 1$, set $v_n(x) = \mathbb{N}_x(1 - \exp -n\langle \mathcal{Z}^D, 1 \rangle)$, $x \in D$. By (A), v_n solves (2) with $g = n$. By Proposition V.9 (iii), $u_1 = \lim \uparrow v_n$ also solves $\Delta u = 4u^2$ in D.

The condition $u_1|_{\partial D} = \infty$ is clear since $u_1 \geq v_n$ and $v_n|_{\partial D} = n$. Finally if v is another nonnegative solution of the problem (3), the comparison principle (Lemma V.7) implies that $v \geq v_n$ for every n and so $v \geq u_1$. □

Proposition 2. *Let D be any open set in $\mathbb{R}^d$ and $u_2(x) = \mathbb{N}_x(\mathcal{R} \cap D^c \neq \emptyset)$ for $x \in D$. Then u_2 is the maximal nonnegative solution of $\Delta u = 4u^2$ in D (in the sense that $u \leq u_2$ for any other nonnegative solution u in D).*

Proof. Recall from Theorem IV.7 that $\mathcal{R}$ is connected $\mathbb{N}_x$ a.e. It follows that we may deal separately with each connected component of D, and thus assume that D is a domain. Then we can easily construct a sequence (D_n) of bounded regular subdomains of D, such that $D = \lim \uparrow D_n$ and $\bar{D}_n \subset D_{n+1}$ for every n. Set

$$v_n(x) = \mathbb{N}_x(\mathcal{Z}^{D_n} \neq 0) , \quad \tilde{v}_n(x) = \mathbb{N}_x(\mathcal{R} \cap D_n^c \neq \emptyset)$$

for $x \in D_n$. By the support property of the exit measure, it is clear that $v_n \leq \tilde{v}_n$. We also claim that $\tilde{v}_{n+1}(x) \leq v_n(x)$ for $x \in D_n$. To verify this, observe that on the event $\{\mathcal{R} \cap D_{n+1}^c \neq \emptyset\}$ there exists a path W_s that hits D_{n+1}^c. For this path W_s, we must have $\tau_{D_n}(W_s) < \zeta_s$, and it follows from the properties of the Brownian snake that

$$A_\sigma^n := \int_0^\sigma dr \, 1_{\{\tau_{D_n}(W_r) < \zeta_r\}} > 0 ,$$

$\mathbb{N}_x$ a.e. on $\{\mathcal{R} \cap D_{n+1}^c \neq \emptyset\}$. However, from the construction of the exit measure in Chapter V, $\langle \mathcal{Z}^{D_n}, 1 \rangle$ is obtained as the local time at level 0 and at time A_σ^n of a reflected Brownian motion started at 0. Since the local time at 0 of a reflected Brownian motion started at 0 immediately becomes (strictly) positive, it follows that $\{\mathcal{R} \cap D_{n+1}^c \neq \emptyset\} \subset \{\mathcal{Z}^{D_n} \neq 0\}$ $\mathbb{N}_x$ a.e., which gives the inequality $\tilde{v}_{n+1}(x) \leq v_n(x)$.

We have then for $x \in D$

$$u_2(x) = \lim_{n \to \infty} \downarrow \tilde{v}_n(x) = \lim_{n \to \infty} \downarrow v_n(x), \tag{4}$$

This follows easily from the fact that the event $\{\mathcal{R} \cap D^c \neq \emptyset\}$ is equal $\mathbb{N}_x$ a.e. to the intersection of the events $\{\mathcal{R} \cap D_n^c \neq \emptyset\}$. By Proposition 1, v_n solves $\Delta u = 4u^2$ in D_n. It then follows from (4) and Proposition V.9 (iii) that u_2 solves $\Delta u = 4u^2$ in D. Finally, if u is another nonnegative solution in D, the comparison principle implies that $u \leq v_n$ in D_n and it follows that $u \leq u_2$. □

Example. Let us apply the previous proposition to compute $\mathbb{N}_x(0 \in \mathcal{R})$ for $x \neq 0$. By rotational invariance and the same scaling argument as in the proof of Proposition V.9 (i), we get $\mathbb{N}_x(0 \in \mathcal{R}) = C|x|^{-2}$ with a nonnegative constant C. On the other hand, by Proposition 2, we know that $u(x) = \mathbb{N}_x(0 \in \mathcal{R})$ solves $\Delta u = 4u^2$ in $\mathbb{R}^d \setminus \{0\}$. A short calculation, using the expression of the Laplacian for a radial function, shows that the only possible values of C are $C = 0$ and $C = 2 - \frac{d}{2}$. Since u is the maximal solution, we conclude that if $d \leq 3$,

$$\mathbb{N}_x(0 \in \mathcal{R}) = \left(2 - \frac{d}{2}\right)|x|^{-2}$$

whereas $\mathbb{N}_x(0 \in \mathcal{R}) = 0$ if $d \geq 4$. In particular, points are polar (in a sense that will be made precise in the next section) if and only if $d \geq 4$.

To conclude this section, let us briefly motivate the results that will be derived below. First note that, if D is bounded and regular (the boundedness is superfluous here), the function u_2 of Proposition 2 also satisfies $u_2|_{\partial D} = +\infty$. This is obvious since $u_2 \geq u_1$. We may ask the following two questions:

1. If D is regular, is it true that $u_1 = u_2$? (uniqueness of the solution with boundary blow-up)

2. For a general domain D, when is it true that $u_2|_{\partial D} = +\infty$? (existence of a solution with boundary blow-up)

We will give a complete answer to question **2** in Section 3. It may well be that the answer to **1** is always yes. We will prove a partial result in this direction in Section 4. Let us however give a simple example of a (nonregular) domain D for which $u_1 \neq u_2$. We let $D = B(0,1)\backslash\{0\}$ be the punctured unit ball in $\mathbb{R}^d$, for $d = 2$ or 3. From Proposition V.3, it is immediate that

$$\mathbb{N}_x\big(\langle \mathcal{Z}^D, 1_{\{0\}}\rangle > 0\big) = 0$$

for every $x \in D$. Hence

$$u_1(x) = \mathbb{N}_x(\mathcal{Z}^D \neq 0) = \mathbb{N}_x\big(\mathcal{Z}^D(\partial B(0,1)) > 0\big)$$

is bounded above on $B(0,1/2)\backslash\{0\}$ by Proposition V.9. On the other hand

$$u_2(x) = \mathbb{N}_x(\mathcal{R} \cap D^c \neq \emptyset) \geq \mathbb{N}_x(0 \in \mathcal{R}) = \left(2 - \frac{d}{2}\right)|x|^{-2}.$$

Clearly, this implies $u_1 \neq u_2$.

2   Polar sets

Definition. *A compact subset K of $\mathbb{R}^d$ is called polar if $\mathbb{N}_x(\mathcal{R} \cap K \neq \emptyset) = 0$ for every $x \in \mathbb{R}^d\backslash K$.*

Because of the relations between the Brownian snake and superprocesses, this is equivalent to the property $\mathbb{P}_\mu(\mathcal{R}^Z \cap K \neq \emptyset) = 0$ for every $\mu \in M_f(\mathbb{R}^d)$ (here Z is under $\mathbb{P}_\mu$ a super-Brownian motion started at μ, and the range $\mathcal{R}^Z$ was defined in Chapter IV).

By applying Proposition 2 to $D = \mathbb{R}^d\backslash K$, we immediately get the following result.

Proposition 3. *K is polar if and only if there exists no nontrivial nonnegative solution of $\Delta u = 4u^2$ in $\mathbb{R}^d\backslash K$.*

In analytic terms, this corresponds to the notion of (interior) *removable singularity* for the partial differential equation $\Delta u = 4u^2$.

If $d \geq 4$, we define the capacity $C_{d-4}(K)$ by the formula

$$C_{d-4}(K) = \left(\inf_{\nu \in \mathcal{M}_1(K)} \iint \nu(dy)\nu(dz) f_d(|y-z|)\right)^{-1}$$

wehere $\mathcal{M}_1(K)$ is the set of all probability measures on K, and

$$f_d(r) = \begin{cases} 1 + \log^+ \frac{1}{r} & \text{if } d = 4\,, \\ r^{4-d} & \text{if } d \geq 5\,. \end{cases}$$

Theorem 4. *If $d \leq 3$, there are no nonempty polar sets. If $d \geq 4$, K is polar if and only if $C_{d-4}(K) = 0$.*

Proof. The case $d \leq 3$ is trivial since we have already seen that points are not polar in dimension $d \leq 3$. From now on, we suppose that $d \geq 4$.

First step. We first prove that K is not polar if $C_{d-4}(K) > 0$. By the definition of $C_{d-4}(K)$, we can find a probability measure ν on K such that

$$\iint \nu(dy)\nu(dz)f_d(|y - z|) < \infty .$$

Let $h : \mathbb{R}^d \to \mathbb{R}_+$ be a radial (i.e. $h(x) = h(y)$ if $|x| = |y|$) continuous function with compact support such that $\int_{\mathbb{R}^d} h(y)dy = 1$. For $\varepsilon > 0$, set $h_\varepsilon(x) = \varepsilon^{-d}h(x/\varepsilon)$.

Recall the notation $\mathcal{J}$ for the "total occupation measure" of the Brownian snake:

$$\langle \mathcal{J}, g \rangle = \int_0^\sigma ds\, g(\hat{W}_s) .$$

By Proposition IV.2, we can compute the first and second moments of $\langle \mathcal{J}, h_\varepsilon * \nu \rangle$ under $\mathbb{N}_x$, $x \in \mathbb{R}^d \backslash K$. If $G(x, y) = G(y - x) = \gamma_d |y - x|^{2-d}$ is the Green function of Brownian motion in $\mathbb{R}^d$, we have first

$$\mathbb{N}_x\left(\langle \mathcal{J}, h_\varepsilon * \nu \rangle\right) = \Pi_x\left(\int_0^\infty dt\, h_\varepsilon * \nu(\xi_t)\right)$$

$$= \int dy\, G(y - x)h_\varepsilon * \nu(y)$$

$$= \int \nu(dz) \int dy\, G(y - x)h_\varepsilon(y - z)$$

and this quantity tends to $\int \nu(dz)G(z - x) > 0$ as ε goes to 0. In particular, there exists a positive constant c_1 (depending on x and K) such that $\mathbb{N}_x\left(\langle \mathcal{J}, h_\varepsilon * \nu \rangle\right) \geq c_1$ for $\varepsilon \in (0, 1]$.

Similarly, Proposition IV.2 allows us to compute the second moment

$$\mathbb{N}_x\left(\langle \mathcal{J}, h_\varepsilon * \nu \rangle^2\right) = 4\,\Pi_x\left(\int_0^\infty dt\, \left(\Pi_{\xi_t}\left(\int_0^\infty dr\, h_\varepsilon * \nu(\xi_r)\right)\right)^2\right)$$

$$= 4 \int da\, G(a - x)\left(\int dy\, G(y - a)h_\varepsilon * \nu(y)\right)^2$$

$$= 4 \iint \nu(dz)\nu(dz') \iint dy\, dy'\, h_\varepsilon(y - z)h_\varepsilon(y' - z')$$

$$\times \int da\, G(a - x)G(y - a)G(y' - a) .$$

By our assumptions on h and the fact that the function G is superharmonic on $\mathbb{R}^d$, we have

$$\int dy\, h_\varepsilon(y - z)G(y - a) \le G(z - a)\,.$$

It follows that

$$N_x\big((\langle \mathcal{J}, h_\varepsilon * \nu\rangle)^2\big) \le 4\iint \nu(dz)\nu(dz')\int da\, G(a - x)G(z - a)G(z' - a).$$

Then Lemma IV.8 gives

$$N_x\big((\langle \mathcal{J}, h_\varepsilon * \nu\rangle)^2\big) \le 4c_2 \iint \nu(dz)\nu(dz')f_d(|z - z'|)$$

with a constant c_2 depending on x and K. From our assumption on ν we get

$$N_x\big((\langle \mathcal{J}, h_\varepsilon * \nu\rangle)^2\big) \le c_3 < \infty\,.$$

By the Cauchy-Schwarz inequality, it follows that

$$N_x\big(\langle \mathcal{J}, h_\varepsilon * \nu\rangle > 0\big) \ge \frac{(N_x(\langle \mathcal{J}, h_\varepsilon * \nu\rangle))^2}{N_x(\langle \mathcal{J}, h_\varepsilon * \nu\rangle^2)} \ge \frac{c_1^2}{c_3} = c_4 > 0\,,$$

where c_4 depends on x and K but not on $\varepsilon \in (0,1]$. Let $r > 0$ be such that $h(y) = 0$ if $|y| \ge r$. Obviously $h_\varepsilon * \nu$ is supported on $K_{r\varepsilon} = \{y, \mathrm{dist}(y, K) \le r\varepsilon\}$. Since $\mathcal{J}$ is supported on $\mathcal{R}$ we get

$$N_x(\mathcal{R} \cap K_{r\varepsilon} \ne \emptyset) \ge c_4$$

and by letting ε go to 0,

$$N_x(\mathcal{R} \cap K \ne \emptyset) \ge c_4\,,$$

which proves that K is not polar.

Second step. We will now verify that K is polar if $C_{d-4}(K) = 0$. The proof is based on an analytic lemma. If $\varphi \in C_0^\infty(\mathbb{R}^d)$, the Sobolev norm $\|\varphi\|_{2,2}$ is

$$\|\varphi\|_{2,2} = \|\varphi\|_2 + \sum_{j=1}^d \left\|\frac{\partial\varphi}{\partial x_j}\right\|_2 + \sum_{j,k=1}^d \left\|\frac{\partial^2\varphi}{\partial x_j \partial x_k}\right\|_2$$

and we introduce the capacity

$$c_{2,2}(K) = \inf\{\|\varphi\|_{2,2}^2 \,; \varphi \in C_0^\infty(\mathbb{R}^d)\,,$$
$$0 \le \varphi \le 1 \text{ and } \varphi = 1 \text{ on a neighborhood of } K\}\,.$$

Lemma 5. *There exist positive constants* α_1, α_2 *such that, for every compact subset H of $[-1,1]^d$, we have*

$$\alpha_1 c_{2,2}(H) \le C_{d-4}(H) \le \alpha_2\, c_{2,2}(H)\,.$$

Proof. Recall the definition of the Bessel kernel G_2 (see [AH] p.10): For $x \in \mathbb{R}^d$,

$$G_2(x) = (4\pi)^{-1} \int_0^\infty t^{-\frac{d}{2}} \exp\left(-\frac{\pi|x|^2}{t} - \frac{t}{4\pi}\right) dt.$$

Set

$$C_{2,2}(H) = \sup_{\mu \in \mathcal{M}_1(H)} \frac{1}{\|G_2 * \mu\|_2^2}\,.$$

As a consequence of Theorem 2.2.7 and Corollary 3.3.4 in [AH], there exists a constant A, depending only on d, such that

$$A^{-1} C_{2,2}(H) \le c_{2,2}(H) \le A\, C_{2,2}(H).$$

It remains to compare $C_{2,2}(H)$ and $C_{d-4}(H)$. To this end, note that

$$\|G_2 * \mu\|_2^2 = \int\int \mu(dy)\, \mu(dy')\, F_d(y - y'),$$

with

$$F_d(y - y') = \int_{\mathbb{R}^d} dz\, G_2(z - y)\, G_2(z - y').$$

Notice that $G_2(z) \sim c|z|^{2-d}$ as $z \to 0$. It is then elementary to verify that, for $y, y' \in [-1,1]^d$, the ratio

$$\frac{F_d(y - y')}{f_d(|y - y'|)}$$

is bounded above and below by positive constants depending only on d. (Compare with Lemma IV.8.) Lemma 5 now follows by comparing the definitions of $C_{d-4}(H)$ and $C_{2,2}(H)$. $\qquad\qquad\square$

Let us complete the proof of Theorem 4. Let K be a compact subset of the unit ball $\bar{B}(0,1)$ (clearly we can restrict our attention to this case). Let $\varphi \in C_0^\infty(\mathbb{R}^d)$ be such that $0 \le \varphi \le 1$ and $\varphi = 1$ on a neighborhood of K. Let $R > 2$ be such that $\varphi(y) = 0$ if $|y| \le R - 1$. Set

$$D_R = (-R, R)^d$$

and $\psi = 1 - \varphi$. Note that $\psi = 1$ on a neighborhood of ∂D_R and $\psi = 0$ on a neighborhood of K. Then set

$$F_R = \bigcup_{m \in \mathbb{Z}^d} (2mR + K).$$

By Proposition 2, the function

$$u_R(x) = \mathbb{N}_x(\mathcal{R} \cap F_R \neq \emptyset), \qquad x \in \mathbb{R}^d \backslash F_R$$

solves $\Delta u = 4u^2$ in $\mathbb{R}^d \backslash F_R$. Furthermore, u_R has period $2R$ in every coordinate direction.

Recalling that $\psi = 0$ on a neighborhood of K, we get after two integrations by parts

$$4 \int_{D_R} \psi(y)^4 u_R(y)^2 dy = \int_{D_R} \psi(y)^4 \Delta u_R(y) dy = \int_{D_R} \Delta(\psi^4)(y) u_R(y) dy \,.$$

In the first integration by parts, we use the fact that $\psi = 1$ on ∂D_R and the periodicity of u_R. In the second one, we use the fact that $\nabla \psi = 0$ on ∂D_R. Then, by expanding $\Delta(\psi^4)$ we arrive at

$$\frac{1}{4} \int_{D_R} |\Delta(\psi^4)| u_R \, dy$$

$$\leq 3 \int_{D_R} \psi^2 |\nabla \psi|^2 u_R \, dy + \int_{D_R} \psi^3 |\Delta \psi| u_R \, dy$$

$$\leq 3 \left(\int_{D_R} \psi^4 u_R^2 dy \right)^{1/2} \left(\int_{D_R} |\nabla \psi|^4 dy \right)^{1/2} + \left(\int_{D_R} \psi^6 u_R^2 dy \right)^{1/2} \left(\int_{D_R} |\Delta \psi|^2 dy \right)^{1/2}$$

$$\leq \left(\int_{D_R} \psi^4 u_R^2 dy \right)^{1/2} \left(3 \left(\int_{D_R} |\nabla \psi|^4 dy \right)^{1/2} + \left(\int_{D_R} |\Delta \psi|^2 dy \right)^{1/2} \right),$$

using the trivial bound $\psi^6 \leq \psi^4$ since $0 \leq \psi \leq 1$. Note that $\Delta \psi = -\Delta \varphi$, $\nabla \psi = -\nabla \varphi$. A simple integration by parts shows that

$$\int |\nabla \varphi|^4 \, dy \leq C \, \|\varphi\|_\infty^2 \, \|\varphi\|_{2,2}^2 = C \, \|\varphi\|_{2,2}^2$$

with a constant C depending only on d. Combining the previous formulas gives

$$4 \int_{D_R} \psi(y)^4 u_R(y)^2 dy = \int_{D_R} \Delta(\psi^4)(y) u_R(y) dy$$

$$\leq C' \, \|\varphi\|_{2,2} \left(\int_{D_R} \psi(y)^4 u_R(y)^2 dy \right)^{1/2},$$

where the constant C' only depends on d. Hence

$$\int_{D_R} \psi(y)^4 u_R(y)^2 dy \leq (C'/4)^2 \|\varphi\|_{2,2}^2 . \tag{5}$$

Now suppose that $C_{d-4}(K) = 0$. By Lemma 5, $c_{2,2}(K) = 0$ and so we can find a sequence of functions $\varphi_n \in C_0^\infty(\mathbb{R}^d)$ such that $0 \leq \varphi_n \leq 1$, $\varphi_n = 1$ on a neighborhood of K and

$$\lim_{n \to \infty} \|\varphi_n\|_{2,2} = 0 .$$

Set $\psi_n = 1 - \varphi_n$ and choose R_n such that $\varphi_n(z) = 0$ if $|z| > R_n - 1$. Obviously we may assume that $R_n \uparrow \infty$. Also set

$$u(x) = \mathbb{N}_x(\mathcal{R} \cap K \neq \emptyset), \qquad x \in \mathbb{R}^d \backslash K$$

and note the trivial bound $u(x) \leq u_{R_n}(x)$ for $x \in D_{R_n} \backslash K$. From the bound (5) applied to φ_n instead of φ, we get

$$\int_{D_{R_n}} \psi_n(y)^4 u(y)^2 dy \leq \int_{D_{R_n}} \psi_n(y)^4 u_{R_n}(y)^2 dy \leq (C'/4)^2 \|\varphi_n\|_{2,2}^2 .$$

Since $\psi_n = 1 - \varphi_n$ and $\|\varphi_n\|_2 \leq \|\varphi_n\|_{2,2} \to 0$, we get from the last bound and Fatou's lemma that

$$\int_{\mathbb{R}^d \backslash K} u(y)^2 dy = 0.$$

It follows that K is polar, which completes the proof of Theorem 4. □

Remark. The second half of the previous proof strongly relies on analytic ingredients. The existence of a probabilistic proof still remains an open problem.

3 Wiener's test for the Brownian snake

In this section we will give a complete answer to a question which was raised in Section 1. Precisely, we will characterize the domains D in $\mathbb{R}^d$ in which there exists a nonnegative solution of $\Delta u = u^2$ that blows up everywhere at the boundary. This characterization will follow from a Wiener-type criterion for the Brownian snake, which is of independent interest.

For $y \in \mathbb{R}^d$ and $0 \leq r < r'$ we denote by $\mathcal{C}(y, r, r')$ the spherical shell

$$\mathcal{C}(y; r, r') = \{z \in \mathbb{R}^d; r \leq |z - y| \leq r'\} .$$

We also define under $\mathbb{N}_x$

$$\mathcal{R}^* = \{\hat{W}_s, 0 < s < \sigma\}.$$

Note that $\mathcal{R} = \mathcal{R}^* \cup \{x\}$, $\mathbb{N}_x$ a.e.

Theorem 6. *Let F be a closed subset of $\mathbb{R}^d$ and let $y \in F$. Then the property $\mathbb{N}_y(\mathcal{R}^* \cap F \neq \emptyset) = \infty$ holds if and only if $d \leq 3$, or $d \geq 4$ and*

$$\sum_{n=1}^{\infty} 2^{n(d-2)} C_{d-4}\big(F \cap \mathcal{C}(y; 2^{-n}, 2^{-n+1})\big) = \infty. \tag{6}$$

From excursion theory, the property $\mathbb{N}_y(\mathcal{R}^* \cap F \neq \emptyset) = \infty$ is equivalent to $\mathbb{P}_y(T_F = 0) = 1$, where

$$T_F = \inf\{s \geq 0; \hat{W}_s \in F \text{ and } \zeta_s > 0\}.$$

Alternatively, if Z is under $\mathbb{P}_{\delta_y}$ a super-Brownian notion started at δ_y, the previous properties are also equivalent to $\mathbb{P}_{\delta_y}(S_F = 0) = 1$, where

$$S_F = \inf\{t > 0; \operatorname{supp} Z_t \cap F \neq \emptyset\}.$$

This essentially follows from the relationship between the Brownian snake and super-Brownian motion, as described in Chapter IV (see [DL] for details). The previous remarks show that Theorem 6 is an analogue of the classical Wiener criterion. In the same way as for the classical Wiener criterion, Theorem 6 has a remarkable analytic counterpart.

Corollary 7. *Let D be a domain in $\mathbb{R}^d$. The problem*

$$\begin{cases} \Delta u = 4u^2 \\ u|_{\partial D} = +\infty \end{cases} \tag{7}$$

has a nonnegative solution if and only if $d \leq 3$, or $d \geq 4$ and (6) holds with $F = D^c$ for every $y \in \partial D$.

In dimension $d \geq 4$, the proof shows more precisely that the existence of a nonnegative solution that blows up at a fixed point y_0 of ∂D is equivalent to condition (6) with $y = y_0$ and $F = D^c$.

The key ingredient of the proof of Theorem 6 is the following proposition, which gives precise estimates on hitting probabilities of compact sets and can be viewed as a reinforcement of Theorem 4. Under $\mathbb{N}_x$, we set

$$M = \sup\{\zeta_s; 0 \leq s \leq \sigma\}.$$

Proposition 8. *Suppose that $d \geq 4$. There exist two positive constants β_1, β_2 such that for every compact subset K of $\mathcal{C}(0; 1, 2)$ and every $x \in B(0, 1/2)$,*

$$\beta_1 C_{d-4}(K) \leq \mathbb{N}_x(\mathcal{R} \cap K \neq \emptyset; 1 < M \leq 2) \leq \mathbb{N}_x(\mathcal{R} \cap K \neq \emptyset) \leq \beta_2 C_{d-4}(K).$$

Proof of the upper bound. By simple translation arguments, it is enough to prove the given upper bound when K is a compact subset of $B(0, 1/2)$ and $|x| > 1$. By Theorem 4, we may assume that $C_{d-4}(K) > 0$. We set

$$u(x) = \mathbb{N}_x(\mathcal{R} \cap K \neq \emptyset), \qquad x \in \mathbb{R}^d \backslash K.$$

From Lemma 5, we can find a function $\varphi \in C_0^\infty(\mathbb{R}^d)$ such that $0 \leq \varphi \leq 1$, $\varphi = 1$ on a neighborhood of K and

$$\|\varphi\|_{2,2}^2 \leq c_0 \, C_{d-4}(K)$$

where the constant c_0 only depends on d. Multiplying φ by a function $h \in C_0^\infty(\mathbb{R}^d)$ such that $h = 1$ on $\bar{B}(0, 1/2)$, $h = 0$ on $\mathbb{R}^d \backslash B(0, 3/4)$ and $0 \leq h \leq 1$, we may assume furthermore that φ vanishes outside $B(0, 3/4)$ (the value of the constant c_0 will be changed but will still depend only on d). As previously, we set $\psi - 1 - \varphi$.

Recall the notation D_R, u_R from the proof of Theorem 4. By (5), we have for every R large enough,

$$\int_{D_R} \psi(y)^4 u_R(y)^2 dy \leq c_1 \|\varphi\|_{2,2}^2$$

with a constant c_1 depending only on d. Observe that $u \leq u_R$ and let R tend to ∞ to get

$$\int_{\mathbb{R}^d} \psi(y)^4 u(y)^2 dy \leq c_1 \|\varphi\|_{2,2}^2. \tag{8}$$

From an intermediate bound of the proof of Theorem 4, we have also

$$\int_{D_R} |\Delta(\psi^4)| u_R \, dy \leq c_2 \|\varphi\|_{2,2}^2,$$

and the same argument gives

$$\int_{\mathbb{R}^d} |\Delta(\psi^4)| u \, dy \leq c_2 \|\varphi\|_{2,2}^2. \tag{9}$$

By applying Itô's formula to ξ under Π_x, we have Π_x a.s.

$$(\psi^4 u)(\xi_t) = u(x) + \int_0^t \nabla(\psi^4 u)(\xi_s) \cdot d\xi_s + \frac{1}{2} \int_0^t \Delta(\psi^4 u)(\xi_s) \, ds.$$

Let $a > |x|$ and $S_a = \inf\{t \geq 0, |\xi_t| \geq a\}$. By applying the optional stopping theorem at $t \wedge S_a$, we get

$$\Pi_x\big((\psi^4 u)(\xi_{t\wedge S_a})\big)$$

$$= u(x) + \frac{1}{2}\Pi_x\Big(\int_0^{t\wedge S_a} \Delta(\psi^4 u)(\xi_s)\, ds\Big)$$

$$= u(x) + \frac{1}{2}\Pi_x\Big(\int_0^{t\wedge S_a} (\psi^4\Delta u + 2\nabla(\psi^4)\cdot\nabla u + \Delta(\psi^4)u)(\xi_s)\, ds\Big).$$

Since $\Delta u = 4u^2 \geq 0$ on $\mathbb{R}^d\backslash K$, we have

$$u(x) \leq \Pi_x\big((\psi^4 u)(\xi_{t\wedge S_a})\big) - \frac{1}{2}\Pi_x\Big(\int_0^{t\wedge S_a} (2\nabla(\psi^4)\cdot\nabla u + \Delta(\psi^4)u)(\xi_s)\, ds\Big).$$

Note that both functions $\nabla(\psi^4)$ and $\Delta(\psi^4)$ vanish outside $B(0, 3/4)$, and $|u(y)|$ tends to 0 as $|y|$ tends to ∞. By letting t, and then a tend to ∞, we get

$$u(x) \leq -\frac{1}{2}\Pi_x\Big(\int_0^\infty (2\nabla(\psi^4)\cdot\nabla u + \Delta(\psi^4)u)(\xi_s)\, ds\Big)$$

$$= -\frac{\gamma_d}{2}\int_{\mathbb{R}^d} (2\nabla(\psi^4)\cdot\nabla u + \Delta(\psi^4)u)(y)\, |y - x|^{2-d}\, dy.$$

$$(10)$$

We will now bound the right side of (10). Since $|x| > 1$ and $\psi = 1$ outside $B(0, 3/4)$, we get

$$\int_{\mathbb{R}^d} |(\Delta(\psi^4)u)(y)|\, |y - x|^{2-d}\, dy \leq 4^{d-2}\int_{\mathbb{R}^d} |(\Delta(\psi^4)u)(y)|\, dy \leq 4^{d-2}c_2\|\varphi\|_{2,2}^2,$$

by (9).

Then consider the other term in the right side of (10). Observe that if $h_d(y) = |y-x|^{2-d}$ we can find a constant c', independent of the choice of x with $|x| > 1$, such that $|\nabla h_d(y)| \leq c'$ for every $y \in B(0, 3/4)$. Then an integration by parts gives

$$\Big|\int_{\mathbb{R}^d} (\nabla(\psi^4)\cdot\nabla u)(y)\, |y - x|^{2-d}\, dy\Big|$$

$$= \Big|\int_{\mathbb{R}^d} (\Delta(\psi^4)u)(y)\, |y - x|^{2-d}\, dy + \int_{\mathbb{R}^d} (u\nabla\psi^4\cdot\nabla h_d)(y)\, dy\Big|$$

$$\leq 4^{d-2}c_2\|\varphi\|_{2,2}^2 + 4c'\int_{\mathbb{R}^d} u\,\psi^3|\nabla\psi|\, dy$$

$$\leq 4^{d-2}c_2\|\varphi\|_{2,2}^2 + 4c'\Big(\int_{\mathbb{R}^d} u^2\psi^4\, dy\Big)^{1/2}\Big(\int_{\mathbb{R}^d} |\nabla\psi|^2\, dy\Big)^{1/2}$$

$$\leq (4^{d-2}c_2 + 4c'c_1^{1/2})\,\|\varphi\|_{2,2}^2$$

by (8). Set $c_3 = \gamma_d(4^{d-2}c_2 + 2c'c_1)$. By substituting the last two bounds in (10), we arrive at

$$\mathbb{N}_x(\mathcal{R} \cap K \neq \emptyset) = u(x) \leq c_3 \|\varphi\|_{2,2}^2 \leq c_0 c_3 C_{d-4}(K),$$

which completes the proof of the upper bound.

Remark. To avoid using the Itô formula, one may try to bound u directly from (8) and a suitable Harnack inequality. However, this only gives the bound $u(x) \leq c C_{d-4}(K)^{1/2}$, which is weaker than the upper bound of Proposition 8.

Proof of the lower bound. We may assume that $C_{d-4}(K) > 0$. Then fix a probability measure ν on K such that

$$\iint \nu(dy)\nu(dy') f_d(|y - y'|) \leq 2C_{d-4}(K)^{-1} .$$

Let $\mathcal{J}, h, h_\varepsilon$ be as in the proof of Theorem 4. Then for every $\varepsilon \in (0,1)$, set

$$U_\varepsilon = 1_{\{M \leq 2\}} \int_0^\sigma ds\, h_\varepsilon * \nu(\hat{W}_s) 1_{\{\zeta_s > 1\}} .$$

Note that $U_\varepsilon \leq \langle \mathcal{J}, h_\varepsilon * \nu \rangle$ and so, by an estimate of the proof of Theorem 4, we have

$$\mathbb{N}_x(U_\varepsilon^2) \leq 4 \iint \nu(dy)\nu(dy') \int da\, G(a - x)G(y - a)G(y' - a)$$

$$\leq c_1 \iint \nu(dy)\nu(dy') f_d(|y - y'|)$$

$$\leq 2c_1 C_{d-4}(K)^{-1} ,$$

where the constant c_1 is independent of ε, x and K provided that $K \subset C(0; 1, 2)$, $x \in B(0, 1/2)$.

We then get a lower bound on $\mathbb{N}_x(U_\varepsilon)$. By the Markov property under $\mathbb{N}_x$, we have

$$\mathbb{N}_x(U_\varepsilon) = \int_0^\infty ds\, \mathbb{N}_x \left(1_{\{\zeta_s > 1\}} h_\varepsilon * \nu(\hat{W}_s) 1_{\{\sup_{r \leq s} \zeta_r \leq 2\}} 1_{\{\sup_{r \geq s} \zeta_r \leq 2\}} \right)$$

$$= \int_0^\infty ds\, \mathbb{N}_x \left(1_{\{\zeta_s > 1\}} h_\varepsilon * \nu(\hat{W}_s) 1_{\{\sup_{r \leq s} \zeta_r \leq 2\}} \mathbb{P}_{W_s} \left(\sup_{r \leq T_0} \zeta_r \leq 2 \right) \right)$$

$$= \int_0^\infty ds\, \mathbb{N}_x \left(1_{\{\zeta_s > 1\}} h_\varepsilon * \nu(\hat{W}_s) 1_{\{\sup_{r \leq s} \zeta_r \leq 2\}} \left(\frac{2 - \zeta_s}{2} \right) \right)$$

$$= \mathbb{N}_x \left(\int_0^\sigma ds\, h_\varepsilon * \nu(\hat{W}_s) 1_{\{\zeta_s > 1, \sup_{r \leq s} \zeta_r \leq 2\}} \left(\frac{2 - \zeta_s}{2} \right) \right) .$$

In the third equality, we used the fact that $(\zeta_s, s \geq 0)$ is under $\mathbb{P}_w$ a reflected Brownian motion started at ζ_w, together with a familiar property of linear Brownian motion. From the invariance of the Itô measure under time-reversal, it immediately follows that $\mathbb{N}_x$ is invariant under the mapping $(W_s, s \geq 0) \rightarrow (W_{(\sigma-s)+}, s \geq 0)$. Using this property, and then the Markov property as previously, we get

$$\mathbb{N}_x(U_\varepsilon) = \mathbb{N}_x\left(\int_0^\sigma ds\, h_\varepsilon * \nu(\hat{W}_s) 1_{\{\zeta_s > 1, \sup_{r \geq s} \zeta_r \leq 2\}} \left(\frac{2 - \zeta_s}{2}\right)\right)$$

$$= \mathbb{N}_x\left(\int_0^\sigma ds\, h_\varepsilon * \nu(\hat{W}_s) 1_{\{1 < \zeta_s \leq 2\}} \left(\frac{2 - \zeta_s}{2}\right)^2\right)$$

By the case $p = 1$ of Proposition IV.2 (ii), we have

$$\mathbb{N}_x(U_\varepsilon) = \Pi_x\left(\int_1^2 dt\, h_\varepsilon * \nu(\xi_t)\left(\frac{2 - t}{2}\right)^2\right)$$

$$= \int \nu(dy) \int dz\, h_\varepsilon(z - y) \int_1^2 dt\, p_t(z - x)\left(\frac{2 - t}{2}\right)^2$$

where $p_t(z)$ is the Brownian transition density. From this last formula it is now clear that we can find a constant $c_2 > 0$ independent of ε, x and K, such that for every $\varepsilon \in (0, 1)$,

$$\mathbb{N}_x(U_\varepsilon) \geq c_2 .$$

By the Cauchy-Schwarz inequality, we obtain

$$\mathbb{N}_x(U_\varepsilon > 0) \geq \frac{(c_2)^2}{c_1} C_{d-4}(K) .$$

Notice that U_ε can be nonzero only on $\{1 < M \leq 2\}$. Letting ε go to 0 as in the proof of Theorem 4 yields the lower bound of Proposition 8.   □

Proof of Theorem 6. Consider first the case $d \leq 3$. It is enough to prove that $\mathbb{N}_y(y \in \mathcal{R}^*) = \infty$ and we can take $y = 0$. A scaling argument shows that $\mathbb{N}_0(0 \in \mathcal{R}^*) = \lambda \mathbb{N}_0(0 \in \mathcal{R}^*)$ for every $\lambda > 0$. Furthermore from the fact that points are not polar when $d \leq 3$ (and using Lemma V.5 for instance) one easily obtains that $\mathbb{N}_0(0 \in \mathcal{R}^*) > 0$. The desired result follows at once.

Suppose then that $d \geq 4$. For simplicity, we treat only the case $d \geq 5$ (the case $d = 4$ is similar with minor modifications). The polarity of points now implies that $\mathbb{N}_y(y \in \mathcal{R}^*) = 0$ (use Lemma V.5). Also notice the scaling property $C_{d-4}(\lambda K) = \lambda^{d-4} C_{d-4}(K)$ for $\lambda > 0$.

Assume first that (6) does not hold. Then, if $F_y = \{z - y; z \in F\}$, we have

$$\mathbb{N}_y(\mathcal{R}^* \cap F \neq \emptyset) = \mathbb{N}_y(\mathcal{R} \cap (F\backslash\{y\}) \neq \emptyset)$$

$$\leq \sum_{n=1}^{\infty} \mathbb{N}_y\left(\mathcal{R} \cap (F \cap C(y; 2^{-n}, 2^{-n+1})) \neq \emptyset\right) + \mathbb{N}_y(\mathcal{R} \cap B(y, 1)^c \neq \emptyset)$$

$$= \sum_{n=1}^{\infty} \mathbb{N}_0\left(\mathcal{R} \cap (F_y \cap C(0; 2^{-n}, 2^{-n+1})) \neq \emptyset\right) + c$$

$$= \sum_{n=1}^{\infty} 2^{2n} \mathbb{N}_0\left(\mathcal{R} \cap (2^n F_y \cap C(0; 1, 2)) \neq \emptyset\right) + c$$

$$\leq \beta_2 \sum_{n=1}^{\infty} 2^{2n} C_{d-4}\left(2^n F_y \cap C(0; 1, 2)\right) + c$$

$$= \beta_2 \sum_{n=1}^{\infty} 2^{n(d-2)} C_{d-4}\left(F \cap C(y; 2^{-n}, 2^{-n+1})\right) + c$$

$$< \infty.$$

We used successively the scaling property of $\mathbb{N}_0$, Proposition 9 and the scaling property of C_{d-4}.

Conversely, assume that (6) holds. Since the sets $\{2^{-2n} < M \leq 2^{-2n+2}\}, n \geq 1$ are disjoint, we get by similar arguments

$$\mathbb{N}_y(\mathcal{R}^* \cap F \neq \emptyset)$$

$$\geq \sum_{n=1}^{\infty} \mathbb{N}_y\left(\mathcal{R} \cap (F \cap C(y; 2^{-n}, 2^{-n+1})) \neq \emptyset; 2^{-2n} < M \leq 2^{-2n+2}\right)$$

$$= \sum_{n=1}^{\infty} 2^{2n} \mathbb{N}_0\left(\mathcal{R} \cap (2^n F_y \cap C(0; 1, 2)) \neq \emptyset; 1 < M \leq 4\right)$$

$$\geq \beta_1 \sum_{n=1}^{\infty} 2^{2n} C_{d-4}\left(2^n F_y \cap C(0; 1, 2)\right)$$

$$= \beta_1 \sum_{n=1}^{\infty} 2^{n(d-2)} C_{d-4}\left(F \cap C(y; 2^{-n}, 2^{-n+1})\right)$$

$$= \infty$$

by (6). This completes the proof of Theorem 6.   □

Proof of Corollary 7. The case $d \leq 3$ is again easy. We verify that the function $u(x) = \mathbb{N}_x(\mathcal{R} \cap D^c \neq \emptyset)$ solves (7). By Proposition 2, we already know that $\Delta u = 4u^2$ in D. Furthermore, for every $y \in \partial D$,

$$u(x) \geq \mathbb{N}_x(y \in \mathcal{R}) = \left(2 - \frac{d}{2}\right)|y - x|^{-2}$$

so that it is obvious that $u_{|\partial D} = \infty$.

Suppose that $d \geq 4$. First assume that condition (6) holds for some $y \in \partial D$ (with $F = D^c$). We will verify that the maximal solution $u(x) = \mathbb{N}_x(\mathcal{R} \cap D^c \neq \emptyset)$ blows up at y. To this end, let $N \geq 1$ be an integer and consider $x \in B(y, 2^{-N-1}) \cap D$. We have

$$u(x) \geq \sum_{n=1}^{N} \mathbb{N}_x \left(\mathcal{R} \cap \left(F \cap C(y; 2^{-n}, 2^{-n+1})\right) \neq \emptyset ; 2^{-2n} < M \leq 2^{-2n+2} \right).$$

By the same arguments as in the proof of Theorem 6, we get for $n \in \{1, \dots, N\}$,

$$\mathbb{N}_x \left(\mathcal{R} \cap \left(F \cap C(y; 2^{-n}, 2^{-n+1})\right) \neq \emptyset ; 2^{-2n} < M \leq 2^{-2n+2} \right)$$
$$\geq \beta_1 2^{n(d-2)} C_{d-4} \left(F \cap C(y; 2^{-n}, 2^{-n+1})\right)$$

and the desired result follows from (6).

Conversely, suppose that for some fixed $y \in \partial D$, the maximal solution $u(x) = \mathbb{N}_x(\mathcal{R} \cap D^c \neq \emptyset)$ satisfies

$$\lim_{D \ni x \to y} u(x) = +\infty.$$

We will then verify that (6) holds, or equivalently that $\mathbb{N}_y(\mathcal{R}^* \cap D^c \neq \emptyset) = \infty$. Fix $N \geq 1$ and choose $\alpha \in (0, 1)$ such that $\mathbb{N}_x(\mathcal{R} \cap D^c \neq \emptyset) \geq N$ for every $x \in B(y, \alpha) \cap D$. Let $\varepsilon > 0$ and $T_\varepsilon = \inf\{s \geq 0, \zeta_s = \varepsilon\}$. By using the strong Markov property at time T_ε and then Lemma V.5, we get

$$\mathbb{N}_y(\mathcal{R}^* \cap D^c \neq \emptyset) \geq \mathbb{N}_y \left(T_\varepsilon < \infty ; \left(1 - \exp -2 \int_0^\varepsilon dt\, \mathbb{N}_{W_{T_\varepsilon}(t)}(\mathcal{R} \cap D^c \neq \emptyset)\right)\right)$$

$$= (2\varepsilon)^{-1} \Pi_y \left(1 - \exp -2 \int_0^\varepsilon dt\, \mathbb{N}_{\xi_t}(\mathcal{R} \cap D^c \neq \emptyset)\right)$$

$$\geq (2\varepsilon)^{-1} \Pi_y \left(1 - \exp -2N \int_0^\varepsilon dt\, 1_{(0,\alpha)}(|\xi_t - y|)\right).$$

In the second line, we used the equality $N_x(T_\varepsilon < \infty) = (2\varepsilon)^{-1}$ and the fact that W_{T_ε} is distributed under $N_y(\cdot \mid T_\varepsilon < \infty)$ as a Brownian path started at y and stopped at time ε. By letting ε go to 0, we get $N_y(\mathcal{R}^* \cap D^c \neq \emptyset) \geq N$, which completes the proof since N was arbitrary. $\square$

Remark. Until now, there exists no analytic approach to Corollary 7. Some partial results have been obtained in particular by Marcus and Véron [MV1].

4   Uniqueness of the solution with boundary blow-up

We now want to address question 1 which was raised in Section 1. We will only obtain a partial result, which (in the special case of equation $\Delta u = u^2$) is still stronger than what has been done by analytic methods.

We assume that $d \geq 2$. If K is a compact subset of $\mathbb{R}^d$, we denote by $C_{d-2}(K)$ the Newtonian (logarithmic if $d = 2$) capacity of K. Recall from Section 1 the notation u_1, resp. u_2, for the minimal, resp. maximal, nonnegative solution with infinite boundary conditions in a bounded regular domain D.

Theorem 9. *Let D be a bounded domain in $\mathbb{R}^d$, $d \geq 2$. Suppose that for every $y \in \partial D$ there exists a constant $c(y) > 0$ such that the inequality*

$$C_{d-2}\big(D^c \cap \bar{B}(y, 2^{-n})\big) \geq c(y) C_{d-2}\big(\bar{B}(y, 2^{-n})\big)$$

holds for all n belonging to a sequence of positive density in $\mathbb{N}$. Then $u_1 = u_2$,

Remark. As an easy application of the classical Wiener test, the assumption of Theorem 9 implies that D is regular.

For the proof of Theorem 9 we need a technical lemma which gives information on the behavior of the paths W_s near their lifetime. We state this lemma but postpone its proof to the end of the section.

Lemma 10. *For every $\delta > 0$ we can choose $A > 1$ large enough so that N_x a.e., for every $s \in (0, \sigma)$,*

$$\lim_{n \to \infty} \inf \frac{1}{n} \mathrm{Card}\Big\{p \leq n; |W_s(t) - \hat{W}_s| \leq A2^{-p}, \forall t \in \big[(\zeta_s - 2^{-2p})^+, \zeta_s\big]\Big\} \geq 1 - \delta.$$

Proof of Theorem 9. Consider the stopping times

$$T = \inf\big\{s \geq 0, \tau(W_s) \leq \zeta_s\big\}, \quad S = \inf\big\{s \geq 0, \tau(W_s) < \zeta_s\big\}.$$

Then $u_2(x) = N_x(T < \infty)$ and, from an argument used in the proof of Proposition 2, it is easy to see that $u_1(x) = N_x(S < \infty)$. We will prove that

$\{T < \infty\} = \{S < \infty\}$ $\mathbb{N}_x$ a.e. by applying the strong Markov property at T (our argument shows in fact that $T = S$, $\mathbb{N}_x$ a.e.).

Recall the notation $T_0 = \inf\{s \geq 0 , \zeta_s = 0\}$. By the strong Markov property,

$$\mathbb{N}_x(S < \infty) = \mathbb{N}_x\left(T < \infty; \mathbb{P}_{W_T}(S < T_0)\right) .$$

Fix a path w such that $\zeta_w = \tau(w)$. It follows from Lemma V.5 that

$$\mathbb{P}_w(S < T_0) = 1 - \exp -2 \int_0^{\zeta_w} dt \, \mathbb{N}_{w(t)}(S < \infty)$$

$$= 1 - \exp -2 \int_0^{\zeta_w} dt \, u_1\big(w(t)\big) .$$

We see that the proof will be complete if we can verify that $\mathbb{N}_x$ a.e. on $\{T < \infty\}$,

$$\int_0^{\zeta_T} dt \, u_1\big(W_T(t)\big) = \infty . \tag{11}$$

From the assumption of Theorem 9 and Lemma 10, we can choose two (random) constants $c_T > 0$ and $A_T < \infty$ such that the properties

(i) $C_{d-2}\big(B(\hat{W}_T, 2^{-p}) \cap D^c\big) \geq c_T C_{d-2}\big(B(\hat{W}_T, 2^{-p})\big)$

(ii) $|W_T(t) - \hat{W}_T| \leq A_T 2^{-p}$, $\forall t \in [\zeta_T - 2^{-2p}, \zeta_T]$

hold for infinitely many $p \geq 1$ with $2^{-2p} \leq \zeta_T$.

Fix one such p and write $a_p = A_T 2^{-p}$ for simplicity. Also set $T_{(p)} = \inf\{s \geq 0, \zeta_s = a_p^2\}$. Then, if $y \in D$, we have

$$u_1(y) = \mathbb{N}_y(S < \infty) \geq \mathbb{N}_y\left(T_{(p)} < \infty, \tau(W_{T_{(p)}}) < a_p^2\right) .$$

Notice that conditionally on $\{T_{(p)} < \infty\}$, $W_{T_{(p)}}$ is a Brownian path in $\mathbb{R}^d$ started at y stopped at time a_p^2. Hence,

$$\mathbb{N}_y\left(\tau(W_{T_{(p)}}) < a_p^2 \mid T_{(p)} < \infty\right) = \Pi_y(\tau < a_p^2).$$

We can now use (i) to get a lower bound on $\Pi_y(\tau < a_p^2)$ when $|y - \hat{W}_T| \leq a_p$. Suppose first that $d \geq 3$, and denote by L the last hitting time of $H :=$ $\bar{B}(\hat{W}_T, 2^{-p}) \cap D^c$ by the Brownian motion ξ. Let e_H be the capacitary measure of H. Then,

$$\Pi_y(\tau < a_p^2) \geq \Pi_y(0 < L < a_p^2) = \int e_H(dz) \int_0^{a_p^2} p_t(z - y) \, dt,$$

where the last equality is the classical formula for the distribution of L (see [PS] p.62). Note that the total mass of e_H is $C_{d-2}(H)$ and that $C_{d-2}(\bar{B}(y, 2^{-n})) = c2^{-n(d-2)}$. Using the previous bound, (i) and a simple estimate for $\int_0^{a_p^2} p_t(z - y)\, dt$, we get the existence of a (random) constant $\alpha_T > 0$ depending only on c_T and A_T such that, if $|y - \hat{W}_T| \le a_p$,

$$N_y\big(\tau(W_{T_{(p)}}) < a_p^2 \mid T_{(p)} < \infty\big) \ge \alpha_T .$$

In dimension $d = 2$, we can get the same bound by using the last exit time for planar Brownian killed at an independent exponential time (we leave details to the reader).

It follows that
$$u_1(y) \ge \alpha_T\, N_y(T_{(p)} < \infty) = \frac{\alpha_T}{2A_T^2 2^{-2p}}.$$

We apply this to $y = W_T(t)$ for $t \in \big[\zeta_T - 2^{-2p}, \zeta_T - 2^{-2p-2}\big]$. We get

$$\int_{\zeta_T - 2^{-2p}}^{\zeta_T - 2^{-2p-2}} dt\, u_1\big(W_T(t)\big) \ge \frac{3\alpha_T}{8A_T^2} =: \beta_T > 0 .$$

Since this bound holds for infinitely many values of p, the proof of (11) is complete.   □

Remark. For every rational $q \ge 0$, we can apply the argument of the previous proof to the stopping time $T_{(q)} = \inf\{s \ge q, \tau(W_s) = \zeta_s\}$ instead of T. It follows that the support of the random measure dL_s^D is exactly equal to the set $\{s \ge 0, \tau(W_s) = \zeta_s\}$. As a consequence, the assumption of Theorem 9 implies that $\operatorname{supp} \mathcal{Z}^D = \mathcal{E}^D$, N_x a.e. The inclusion $\operatorname{supp} \mathcal{Z}^D \subset \mathcal{E}^D$ is always true but the converse may be false as we see from the example of the punctured unit ball in $\mathbb{R}^d$, $d = 2$ or 3.

Proof of Lemma 10. For every stopped path w, every $A > 0$ and every integer $n \ge 1$, set

$$F_n^A(w) = \operatorname{Card}\big\{p \in \{1, ..., n\}, |w(t) - \hat{w}| \le A2^{-p}, \ \forall t \in [(\zeta_w - 2^{-2p})^+, \zeta_w]\big\}.$$

We first state a simple large deviation estimate for standard Brownian motion, whose easy proof is left to the reader. The notation $\xi_{[0,t]}$ stands for the stopped path $(\xi_r, 0 \le r \le t)$.

Lemma 11. *Let $\delta > 0$ and $\lambda > 0$. We can choose $A > 0$ large enough so that, for every $n \ge 1$ and every $t > 0$,*

$$\Pi_0\big(F_n^A(\xi_{[0,t]}) < (1 - \delta)n\big) \le e^{-\lambda n}.$$

Then, for every $n \geq 1$, introduce the stopping times σ_i^n defined inductively as follows:

$$\sigma_0^n = 0, \qquad \sigma_{i+1}^n = \inf\{t \geq \sigma_i^n, |\zeta_s - \zeta_{\sigma_i^n}| = 2^{-2n}\}.$$

Note that $\sigma_i^n < \infty$ if and only if $\sigma_i^n < L_n$, where $L_n < \infty$, $\mathbb{N}_x$ a.e. Under $\mathbb{N}_x(\cdot \mid \sigma_1^n < \infty)$, the sequence $(2^{2n}\zeta_{\sigma_i^n}, 0 \leq i \leq L_n)$ is distributed as the positive excursion of simple random walk. By a well-known result, we have for every $k \geq 1$,

$$\mathbb{N}_x\Big(\sum_{i=1}^{\infty} 1_{\{\sigma_i^n < \infty,\ \zeta_{\sigma_i^n} = k2^{-2n}\}}\Big) = 2\,\mathbb{N}_x(\sigma_1^n < \infty) = 2^{2n}.$$

On the other hand, under $\mathbb{N}_x(\cdot \mid \sigma_i^n < \infty)$ and conditionally on $\zeta_{\sigma_i^n}$, $W_{\sigma_i^n}$ is distributed as a Brownian path in $\mathbb{R}^d$ started at x and with lifetime $\zeta_{\sigma_i^n}$ (this is so because σ_i^n is a measurable function of the lifetime process).

Let $\delta > 0$, $\lambda > 4$ and choose A as in Lemma 11. By combining the previous observations with this lemma, we get for every integer $M \geq 1$

$$\mathbb{N}_x\Big(\sum_{i=1}^{\infty} 1_{\{\sigma_i^n < \infty,\ \zeta_{\sigma_i^n} \leq M\}} 1_{\{F_n^A(W_{\sigma_i^n}) < (1-\delta)n\}}\Big) \leq M 2^{2n}\, 2^{2n}\, e^{-\lambda n}.$$

From the Borel-Cantelli lemma, we conclude that $\mathbb{N}_x$ a.e. there exists an integer $N_0(\omega)$ such that for $n \geq N_0(\omega)$,

$$F_n^A(W_{\sigma_i^n}) \geq (1 - \delta)n, \qquad \forall i \in \{1, \dots, L_n\}. \tag{12}$$

To complete the proof, we need to "interpolate" between σ_i^n and σ_{i+1}^n. First note that the law under $\mathbb{N}_x(\cdot \mid \sigma_i^n < \infty)$ of $\sigma_{i+1}^n - \sigma_i^n$ is the law of the first exit time from $[-2^{-2n}, 2^{-2n}]$ of a linear Brownian motion started at 0. By standard estimates, it follows that, for every $\eta > 0$, we have for n sufficiently large,

$$\sigma_{i+1}^n - \sigma_i^n \leq 2^{-4n(1-\eta)}, \forall i \in \{0, \dots, L_n - 1\}.$$

On the other hand, we also know from Chapter IV that the mapping $s \longrightarrow W_s$ is Hölder continuous with exponent $\frac{1}{4} - \varepsilon$, for every $\varepsilon > 0$. By combining this property with the previous estimate, we get that $\mathbb{N}_x$ a.e. for every $\varepsilon > 0$, there exists an integer $N_1(\omega)$ such that for every $n \geq N_1(\omega)$, every $i \in \{0, \dots, L_n-1\}$ and $s \in [\sigma_i^n, \sigma_{i+1}^n]$,

$$|W_s(t \wedge \zeta_s) - W_{\sigma_i^n}(t \wedge \zeta_{\sigma_i^n})| \leq 2^{-n(1-\varepsilon)}, \qquad \forall t \geq 0$$
$$|\hat{W}_s - \hat{W}_{\sigma_i^n}| \leq 2^{-n(1-\varepsilon)}.$$

This implies for $t \in [0, \zeta_s]$

$$|W_s(t) - \hat{W}_s| \leq |W_s(t) - W_{\sigma_i^n}(t \wedge \zeta_{\sigma_i^n})| + |W_{\sigma_i^n}(t \wedge \zeta_{\sigma_i^n}) - \hat{W}_{\sigma_i^n}| + |\hat{W}_{\sigma_i^n} - \hat{W}_s|$$
$$\leq 2 \cdot 2^{-n(1-\varepsilon)} + |W_{\sigma_i^n}(t \wedge \zeta_{\sigma_i^n}) - \hat{W}_{\sigma_i^n}|.$$

Thanks to this bound and the trivial inequality $|\zeta_s - \zeta_{\sigma_i^n}| \leq 2^{-2n}$ for $s \in [\sigma_i^n, \sigma_{i+1}^n]$, it is then elementary to verify that $\mathbb{N}_x$ a.e. for all n sufficiently large, we have for every $i \in \{0, \ldots, L_n - 1\}$ and every $s \in [\sigma_i^n, \sigma_{i+1}^n]$,

$$F_n^{2(A+1)}(W_s) \geq F_n^A(W_{\sigma_i^n}) - \varepsilon n - 2.$$

From (12) we have then $\mathbb{N}_x$ a.e. for every $s \in [0, \sigma]$,

$$\liminf_{n \to \infty} \frac{1}{n} F_n^{2(A+1)}(W_s) \geq 1 - \delta - \varepsilon.$$

This completes the proof since δ and ε were arbitrary.   □

Chapter VII
The Probabilistic Representation
of Positive Solutions

In this chapter, we address the general problem of providing a probabilistic classification of positive solutions to the partial differential equation $\Delta u = u^2$ in a smooth domain. We give a complete solution to this problem in the case of the planar unit disk. Precisely, we show that solutions are in one-to-correspondence with their traces, where the trace of a solution consists of a compact subset of the boundary and a Radon measure on the complement of this compact subset in the boundary. Furthermore, we give an explicit probabilistic formula for the solution associated with a given trace. At the end of the chapter, we discuss extensions to higher dimensions or more general equations.

1 Singular solutions and boundary polar sets

In this chapter, the spatial motion ξ is again Brownian motion in $\mathbb{R}^d$, and D is a bounded domain of class C^2 in $\mathbb{R}^d$. In Chapter V, we considered solutions of $\Delta u = 4u^2$ in D which are of the form

$$u_g(x) = \mathbb{N}_x(1 - \exp -\langle \mathcal{Z}^D, g \rangle) , \qquad x \in D,$$

where $g \in \mathcal{B}_{b+}(\partial D)$. Our first proposition provides another class of solutions. Recall from Chapter V the notation $\mathcal{E}^D$ for the set of exit points of the paths W_s from D.

Proposition 1. *Let K be a compact subset of ∂D. Then the function*

$$u_K(x) = \mathbb{N}_x(\mathcal{E}^D \cap K \neq \emptyset) , \qquad x \in D,$$

is the maximal nonnegative solution of the problem

$$\begin{cases} \Delta u = 4u^2 , & in \ D , \\ u_{|\partial D \setminus K} = 0 . \end{cases}$$

Remark. When $K = \partial D$, we recover a special case of Proposition VI.2.

Proof. We first verify that u_K solves $\Delta u = 4u^2$ in D. Fix $\varepsilon > 0$ and set $K_\varepsilon = \{y \in \partial D, \mathrm{dist}(y, K) < \varepsilon\}$. By Theorem V.6, we know that for every n the function

$$u_n^\varepsilon(x) = \mathbb{N}_x(1 - \exp-\langle \mathcal{Z}^D, n 1_{K_\varepsilon}\rangle), \qquad x \in D$$

solves $\Delta u = 4u^2$ in D. Clearly, $\mathcal{Z}^D(K_\varepsilon) = 0$ a.e. on the event $\{\mathcal{E}^D \cap K_\varepsilon = \emptyset\}$ and on the other hand, by a remark following the proof of Theorem VI.9, we have also $\mathcal{Z}^D(K_\varepsilon) > 0$ a.e. on the event $\{\mathcal{E}^D \cap K_\varepsilon \neq \emptyset\}$. It follows that

$$\lim_{n \to \infty} u_n^\varepsilon(x) = \mathbb{N}_x(\mathcal{E}^D \cap K_\varepsilon \neq \emptyset) =: u^\varepsilon(x).$$

By Proposition V.9 (iii), we get that that u^ε solves $\Delta u = 4u^2$ in D, and then that $u_K = \lim \downarrow u^\varepsilon$ is also a solution.

We then verify that $u_{K|\partial D \setminus K} = 0$. Fix $y \in \partial D \setminus K$ and choose $\delta > 0$ such that $\mathrm{dist}(y, K) > 2\delta$. For every path $w \in \mathcal{W}$, set $\tau_{(\delta)}(w) = \inf\{t \geq 0, |w(t) - w(0)| \geq \delta\}$. Clearly, if $x \in D$ and $|x - y| < \delta$, we have

$$\mathbb{N}_x(\mathcal{E}^D \cap K \neq \emptyset) \leq \mathbb{N}_x(\exists s \geq 0 : \tau_{(\delta)}(W_s) < \tau(W_s)),$$

where τ is as usual the first exit time from D. Therefore, it is enough to verify that the quantity in the right side goes to 0 as $x \to y$, $x \in D$. To this end, write for every $\alpha > 0$

$$\mathbb{N}_x(\exists s \geq 0 : \tau_{(\delta)}(W_s) < \tau(W_s))$$
$$\leq \mathbb{N}_x(\exists s \in [0, \alpha] \cup [(\sigma - \alpha)^+, \sigma] : \tau_{(\delta)}(W_s) < \infty)$$
$$+ \mathbb{N}_x(\sigma \geq 2\alpha ; \exists s \in [\alpha, \sigma - \alpha] : \tau_{(\delta)}(W_s) < \tau(W_s)).$$

The first term in the right side does not depend on x and goes to 0 as $\alpha \to 0$ by dominated convergence (recall that $\mathbb{N}_x(\exists s \geq 0 : \tau_{(\delta)}(W_s) < \infty) < \infty$ by Proposition V.9 (i)). Thus it suffices to verify that for every fixed $\alpha > 0$ the second term tends to 0 as $x \to y$, $x \in D$. To this end, use the snake property to observe that the paths W_s, $s \in [\alpha, \sigma - \alpha]$ all coincide up to time $m_\alpha = \inf_{[\alpha, \sigma - \alpha]} \zeta_s > 0$. By conditioning first with respect to the lifetime process, we get

$$\mathbb{N}_x(\sigma \geq 2\alpha; \exists s \in [\alpha, \sigma - \alpha] : \tau_{(\delta)}(W_s) < \tau(W_s)) \leq \mathbb{N}_x(\sigma \geq 2\alpha, \Pi_x(\tau_{(\delta)} \wedge m_\alpha < \tau))$$

and the right side of the last formula goes to 0 as $x \to y$ by dominated convergence (the smoothness of D implies that $\Pi_x(\tau > \varepsilon) \to 0$ as $x \to y$, for every $\varepsilon > 0$). This completes the proof of the property $u_{K|\partial D \setminus K} = 0$.

It remains to verify that u_K is the maximal solution of the problem stated in Proposition 1. Let x_0 be a fixed point in D. For every $\delta > 0$, denote by $D_{(\delta)}$ the connected component of the open set

$$\{x \in D, \text{dist}(x, K) > \delta\}$$

that contains x_0 (this makes sense if δ is small enough). Note that $D_{(\delta)}$ is a regular domain. Also set

$$U_{(\delta)} = \partial D_{(\delta)} \backslash \partial D,$$

and, for every $x \in D_{(\delta)}$,

$$u_{(\delta)}(x) = \mathbb{N}_x(\mathcal{Z}^{D_{(\delta)}}(U_{(\delta)}) \neq 0).$$

Writing

$$u_{(\delta)}(x) = \lim_{n \to \infty} \uparrow \mathbb{N}_x \left(1 - \exp{-n} \int \mathcal{Z}^{D_{(\delta)}}(dy) \, \text{dist}(y, \partial D_{(\delta)} \backslash U_{(\delta)}) \right),$$

and arguing as in the proof of Proposition VI.1, we easily get that $u_{(\delta)}$ solves $\Delta u = 4u^2$ in $D_{(\delta)}$ and $u_{(\delta)|U_{(\delta)}} = \infty$.

Finally, let v be another nonnegative solution of the problem stated in Proposition 1. The comparison principle (Lemma V.7) implies that $v \leq u_{(\delta)}$ on $D_{(\delta)}$. However,

$$u_{(\delta)}(x) \leq \mathbb{N}_x(\mathcal{E}^{D_{(\delta)}} \cap U_{(\delta)} \neq \emptyset)$$

and it is easy to check that, for every $x \in D$,

$$\lim_{\delta \to 0} \mathbb{N}_x(\mathcal{E}^{D_{(\delta)}} \cap U_{(\delta)} \neq \emptyset) = u_K(x).$$

The claim $v \leq u_K$ follows, and this completes the proof of Proposition 1.   □

An informal guess is that any nonnegative solution of $\Delta u = 4u^2$ in D could be obtained as a "mixture" of (generalized forms of) solutions of the type u_g and u_K. In dimension $d = 2$, this guess is correct and a precise statement will be given in Section 3 below in the special case of the unit disk (see Theorem 5). In higher dimensions, the problem becomes more complicated and is still the subject of active research (see the discussion in Section 4).

To understand why dimension two (or one) is different, let us introduce the notion of boundary polar set.

Definition. *A compact subset K of ∂D is called boundary polar if $\mathbb{N}_x(\mathcal{E}^D \cap K \neq \emptyset) = 0$ for every $x \in D$.*

Then K is boundary polar if and only if the problem stated in Proposition 1 has no nontrivial nonnegative solution. In this sense, we may say that K is a boundary removable singularity for $\Delta u = u^2$ in D.

If $d \geq 3$, we define the capacity $C_{d-3}(K)$ by the formula

$$C_{d-3}(K) = \left(\inf_{\nu \in \mathcal{M}_1(K)} \int\int \nu(dy)\nu(dz)g_d(|y - z|) \right)^{-1}$$

where

$$g_d(r) = \begin{cases} 1 + \log^+ \frac{1}{r} & \text{if } d = 3 , \\ r^{3-d} & \text{if } d \geq 4 . \end{cases}$$

We state without proof the following theorem, which will not be used in the rest of this chapter.

Theorem 2. *If $d \leq 2$, there are no nonempty boundary polar sets. If $d \geq 3$, K is boundary polar if and only if $C_{d-3}(K) = 0$.*

For $d = 2$, it is enough to verify that singletons are not boundary polar. The proof of this fact is relatively easy by estimating the first and second moments of the exit measure evaluated on a small ball on the boundary (cf (2) and (3) below), and then using the Cauchy-Schwarz inequality to get a lower bound on the "probability" that $\mathcal{E}^D$ intersects this ball. In the case of the unit disk, this can also be viewed as a by-product of the (much stronger) Theorem 5.

A proof of Theorem 2 in dimension $d \geq 3$ can be given along the lines of the characterization of polar sets in Chapter VI. We refer to [L9] and [DK2]. The latter paper deals with more general equations of the type $\Delta u = u^\alpha$, $1 < \alpha \leq 2$.

2   Some properties of the exit measure from the unit disk

In this section and the next one, we restrict our attention to the case when $d = 2$ and D is the open unit disk. However, all results can be extended to a domain D of class C^2 in the plane (see [L11]). We often identify $\mathbb{R}^2$ with the complex plane $\mathbb{C}$, and the boundary ∂D of D with $\mathbb{T} = \mathbb{R}/2\pi\mathbb{Z}$. The Lebesgue measure on ∂D is denoted by $\theta(dz)$.

Let $G_D(x, y)$, resp. $P_D(x, z)$ be the Green function, resp. the Poisson kernel of D. Note the explicit expressions:

$$G_D(x, y) = \frac{1}{\pi} \log \frac{|\tilde{y} - x| \, |y|}{|y - x|} , \quad x, y \in D$$

$$P_D(x, z) = \frac{1}{2\pi} \frac{1 - |x|^2}{|z - x|^2} , \qquad x \in D, z \in \partial D,$$

$$(1)$$

where $\tilde{y} = y/|y|^2$.

We first propose to derive certain properties of the exit measure $\mathcal{Z}^D$. Let $g \in \mathcal{B}_{b+}(\partial D)$. By the cases $p = 1$ and $p = 2$ of Theorem V.10, we have for $x \in D$

$$N_x(\langle \mathcal{Z}^D, g \rangle) = \int_{\partial D} \theta(dy)\, P_D(x, y)\, g(y), \tag{2}$$

and

$$N_x(\langle \mathcal{Z}^D, g \rangle^2) = 4 \int_D dy\, G_D(x, x') \Big(\int_{\partial D} \theta(dy)\, P_D(x', y)\, g(y) \Big)^2. \tag{3}$$

Proposition 3. *Let $x \in D$. Then N_x a.e., the measure $\mathcal{Z}^D$ has a continuous density $z_D(y)$, $y \in \partial D$ with respect to $\theta(dz)$. Furthermore,*

$$N_x\big(z_D(y)\big) = P_D(x, y)\,, \qquad N_x\big(z_D(y)^2\big) = 4 \int_D da\, G_D(x, a)\, P_D(a, y)^2.$$

Proof. For every $\varepsilon > 0$ and $y \in \partial D$, set

$$Z_\varepsilon(y) = (2\varepsilon)^{-1} \mathcal{Z}^D(N_\varepsilon(y)),$$

where $N_\varepsilon(y) = \{z \in \partial D, |z - y| < \varepsilon\}$. By (3) and a polarization argument, we have for every $y, y' \in \partial D$,

$$N_x\big(Z_\varepsilon(y) Z_\varepsilon(y')\big) = (\varepsilon \varepsilon')^{-1} \int_{N_\varepsilon(y)^2} \theta(dz)\theta(dy')\, \psi_x(z, z'),$$

where

$$\psi_x(z, z') = \int_D da\, G_D(x, a)\, P_D(a, z)\, P_D(a, z').$$

From the explicit expressions for G_D and P_D (which yield the easy bounds $P_D(a, z) \leq C|a - z|^{-1}$, $G_D(x, a) \leq C(x)\mathrm{dist}(a, \partial D)$) it is a simple exercise to verify that the function ψ_x is bounded and continuous over $\partial D \times \partial D$. It follows that

$$\lim_{\varepsilon, \varepsilon' \to 0} N_x\big(Z_\varepsilon(y) Z_{\varepsilon'}(y')\big) = 4\, \psi_x(y, y'),$$

and the convergence is uniform when y and y' vary over ∂D. By the Cauchy criterion, it follows that $Z_\varepsilon(y)$ converges in $L^2(N_x)$ as $\varepsilon \to 0$, uniformly in $y \in \partial D$. Hence, we can choose a sequence ε_n decreasing to 0 so that $Z_{\varepsilon_n}(y)$ converges N_x a.e. for every $y \in \partial D$. The process $(z_D(y), y \in \partial D)$ defined by

$$z_D(y) = \lim_{n \to \infty} Z_{\varepsilon_n}(y)$$

$(z_D(y) = 0$ if the limit does not exist) is measurable. Furthermore, for $g \in C_{b+}(\partial D)$,

$$\int \theta(dy) \, z_D(y) \, g(y) = \lim_{\varepsilon \to 0} \int \theta(dy) \, Z_\varepsilon(y) \, g(y),$$

in $L^2(\mathbb{N}_x)$, and on the other hand

$$\int \theta(dy) \, Z_\varepsilon(y) \, g(y) = \int \mathcal{Z}^D(dy) \, (2\varepsilon)^{-1} \int_{N_\varepsilon(y)} \theta(dz) \, g(z) \xrightarrow[\varepsilon \to 0]{} \langle \mathcal{Z}^D, g \rangle.$$

Thus $\mathcal{Z}^D(dy) = z_D(y)\theta(dy)$, $\mathbb{N}_x$ a.e.

It remains to verify that the process $(z_D(y), y \in \partial D)$ has a continuous modification. The previous arguments immediately give the formula

$$\mathbb{N}_x \big((z_D(y) - z_D(y'))^2 \big) = 4\big(\psi_x(y, y) - 2\psi_x(y, y') + \psi_x(y', y')\big)$$

$$= 4 \int_D da \, G_D(x, a) \, \big(P_D(a, y) - P_D(a, y')\big)^2.$$

From the explicit expressions for G_D and P_D, it is then a simple matter to derive the bound

$$\mathbb{N}_x \big((z_D(y) - z_D(y'))^2 \big) \le C(x) \, |y - y'|,$$

with a finite constant $C(x)$ depending only on x. Unfortunately, this is not quite sufficient for the existence of a continuous modification. One way out is to estimate fourth moments. This is a bit more technical and we only sketch the proof, referring to [L11] for a detailed argument. The fourth moment formula for the exit measure (Theorem V.10) leads to

$$\mathbb{N}_x \big((z_D(y) - z_D(y'))^4 \big) = 8.4! \, (F_1(y, y') + 4 \, F_2(y, y')),$$

where

$$F_1(y,y') = \int_{D^3} da_1 da_2 da_3 G_D(x, a_1) G_D(a_1, a_2) G_D(a_1, a_3)$$
$$\times (P_D(a_2, y) - P_D(a_2, y'))^2 (P_D(a_3, y) - P_D(a_3, y'))^2,$$

$$F_2(y, y') = \int_{D^3} da_1 da_2 da_3 G_D(x, a_1) G_D(a_1, a_2) G_D(a_2, a_3)$$
$$\times (P_D(a_1, y) - P_D(a_1, y'))(P_D(a_2, y) - P_D(a_2, y'))(P_D(a_3, y) - P_D(a_3, y'))^2.$$

From this explicit expression and after some lengthy calculations, one arrives at the bound

$$\mathbb{N}_x((z_D(y) - z_D(y'))^4) \le C \, |y - y'|^2, \tag{4}$$

with a finite constant C that can be chosen independently of x provided that x varies in a compact subset of D. The existence of a continuous modification of the process $(z_D(y), y \in \partial D)$ is then a consequence of the classical Kolmogorov lemma.

Finally, the formula $\mathbb{N}_x(z_D(y)^2) = 4\psi_x(y, y)$ is immediate from our approach and the first moment formula for $z_D(y)$ is easy from (2).   □

If D is replaced by $D_r = \{y \in \mathbb{R}^2, |y| < r\}$, the same method (or a scaling argument) shows that, for every $x \in D_r$, the exit measure $\mathcal{Z}^{D_r}$ has ($\mathbb{N}_x$ a.e.) a continuous density z_{D_r} with respect to the Lebesgue measure θ_r on ∂D_r. We will need a weak continuity property of z_{D_r} as a function of $r > 0$. To simplify notation, we write $z_D(r, y) = z_{D_r}(ry)$ for $y \in \partial D$ and $r > 0$.

Lemma 4. *There exists a strictly increasing sequence (r_n) converging to 1 such that, for every $x \in D$,*

$$\lim_{n \to \infty} \left(\sup_{y \in \partial D} |z_D(r_n, y) - z_D(y)| \right) = 0, \qquad \mathbb{N}_x \ a.e.$$

Proof. Let $r \in (0, 1]$ and let τ_r denote the exit time from D_r. From the method of proof of Theorem V.10, it is easy to get for every $x \in D_r$, $g \in \mathcal{B}_{b+}(\partial D)$ and $g' \in \mathcal{B}_{b+}(D_r)$,

$$\mathbb{N}_x(\langle \mathcal{Z}^D, g \rangle \langle \mathcal{Z}^{D_r}, g' \rangle) = 4 \, \Pi_x \left(\int_0^{\tau_r} dt \, \Pi_{\xi_t} \left(g(\xi_\tau) g'(\xi_{\tau_r}) \right) \right)$$

$$= 4 \int_{D_r} da G_{D_r}(x, a) \int \theta(dy) P_D(a, y) g(y) \int \theta_r(dy') P_{D_r}(a, y') g'(y'),$$

where $P_{D_r}(a, y) = r^{-1} P_D(\frac{a}{r}, \frac{y}{r})$ and $G_{D_r}(a, y) = G_D(\frac{a}{r}, \frac{y}{r})$ are respectively the Poisson kernel and the Green function of D_r. From the L^2 construction of z_D, it follows that, for every $y, y' \in \partial D$,

$$\mathbb{N}_x(z_D(y) z_D(r, y')) = 4 \int_{D_r} da \, G_{D_r}(x, a) \, P_D(a, y) \, P_{D_r}(a, ry').$$

Hence,

$$\mathbb{N}_x((z_D(y) - z_D(r, y'))^2) = 4 \left(\int_{D \setminus D_r} da \, G_D(x, a) \, P_D(a, y)^2 \right.$$

$$+ \int_{D_r} da \, (G_D(x, a) - G_{D_r}(x, a)) P_D(a, y)^2$$

$$\left. + \int_{D_r} da \, G_{D_r}(x, a) \left(P_D(a, y) - P_{D_r}(a, ry) \right)^2 \right).$$

From this explicit formula and elementary estimates (using the previously mentioned bounds on P_D and G_D), one easily obtains that

$$\lim_{r \uparrow 1} \left(\sup_{y \in \partial D} \mathbb{N}_x \left((z_D(y) - z_D(r,y))^2 \right) \right) = 0 , \tag{5}$$

and the convergence is uniform when x varies over compact subsets of D.

Let K be a compact subset of D. By (4) and a scaling argument, the bound

$$\mathbb{N}_x \left((z_D(r,y) - z_D(r,y'))^4 \right) \leq C_K \, |y - y'|^2$$

holds for every $x \in K$ and r sufficiently close to 1, with a constant C_K depending only on K. Then, for every $n \geq 1, p \in \mathbb{Z}/2^n \mathbb{Z}$, set $y_p^n = \exp(2i\pi p 2^{-n}) \in \partial D$. Let $\gamma > 0$. The previous bound gives for $k \geq 1$ and $p \in \mathbb{Z}/2^k \mathbb{Z}$,

$$\mathbb{N}_x \left(|z_D(r, y_p^k) - z_D(r, y_{p+1}^k)| > 2^{-\gamma k} \right) \leq C' 2^{4\gamma k} 2^{-2k},$$

where $C' = 4\pi^2 C_K$. We take $\gamma = 1/8$, sum over p and then over $k \geq n$ to get

$$\mathbb{N}_x (\exists k \geq n, \exists p : |z_D(r, y_p^k) - z_D(r, y_{p+1}^k)| > 2^{-k/8}) \leq C'' 2^{-n/2}, \tag{6}$$

where $C'' = C'/(1 - 2^{-1/2})$. Denote by $E_n(r)$ the event

$$E_n(r) = \{ \forall k \geq n, \forall p, |z_D(r, y_p^k) - z_D(r, y_{p+1}^k)| \leq 2^{-k/8} \}.$$

The classical chaining argument of the proof of the Kolmogorov lemma shows that on the set $E_n(r)$ we have for every $y, y' \in \partial D$ such that $|y - y'| \leq 2\pi 2^{-n}$,

$$|z_D(r, y) - z_D(r, y')| \leq c |y - y'|^{1/8}, \tag{7}$$

for some universal constant c. On the other hand, by (5), we may choose a sequence (r_n) that increases to 1 sufficiently fast so that, for every $x \in K$,

$$\mathbb{N}_x \left(\sum_{\{n, r_n > |x|\}} \sum_{p=0}^{2^n - 1} \left(z_D(r_n, y_p^n) - z_D(y_p^n) \right)^2 \right) < \infty.$$

The sequence (r_n) can be chosen independently of the compact set K by extracting a diagonal subsequence. It follows that

$$\lim_{n \to \infty} \sup_{0 \leq p \leq 2^n - 1} |z_D(r_n, y_p^n) - z_D(y_p^n)| = 0, \ \mathbb{N}_x \text{ a.e.}$$

for every $x \in D$. To complete the proof, notice that $\sum_n \mathbb{N}_x(E_n(r_n)^c) < \infty$ by (6). Therefore, the bound (7) holds with $r = r_n$ for all n sufficiently large, $\mathbb{N}_x$ a.e. Lemma 4 follows by writing

$$\sup_{y \in \partial D} |z_D(r_n, y) - z_D(y)| \leq \sup_{0 \leq p \leq 2^n - 1} |z_D(r_n, y_p^n) - z_D(y_p^n)|$$

$$+ \sup_{|y - y'| \leq 2\pi 2^{-n}} |z_D(r_n, y) - z_D(r_n, y')| + \sup_{|y - y'| \leq 2\pi 2^{-n}} |z_D(y) - z_D(y')|.$$

$\square$

3   The representation theorem

We are now ready to state and prove the main result of this chapter. Recall the notation $N_\varepsilon(y) = \{z \in \partial D, |z - y| < \varepsilon\}$.

Theorem 5. *Nonnegative solutions of equation $\Delta u = 4\,u^2$ in D are in one-to-one correspondence with pairs (K, ν), where K is a compact subset of ∂D and ν is a Radon measure on $\partial D \backslash K$.*
In this correspondence, the pair (K, ν) is determined from u by the formulas

$$K = \{y \in \partial D, \lim_{r\uparrow 1, r < 1} \int_{N_\varepsilon(y)} \theta(dz)\, u(rz) = \infty, \quad \text{for every } \varepsilon > 0\}, \quad (8)$$

and, for every $g \in C_0(\partial D \backslash K)$,

$$\langle \nu, g \rangle = \lim_{r\uparrow 1, r < 1} \int_{\partial D \backslash K} \theta(dz)\, u(rz)\, g(z). \quad (9)$$

Conversely, for every $x \in D$,

$$u(x) = \mathbb{N}_x(\mathcal{E}^D \cap K \neq \emptyset) + \mathbb{N}_x\left(1_{\{\mathcal{E}^D \cap K = \emptyset\}}\left(1 - \exp - \int \nu(dy)\, z_D(y)\right)\right). \quad (10)$$

The pair (K, ν) is called the trace of the solution u. Roughly speaking, K corresponds to a singular boundary set for u, and ν to the boundary value of u on $\partial D \backslash K$. As special cases of formula (10), we get $u = u_K$ when $\nu = 0$, and $u = u_g$ when $K = \emptyset$ and $\nu(dy) = g(y)\theta(dy)$. In this sense, the general form of a solution is a mixture of the formulas for u_K and u_g.

Proof. Step 1. We first verify that, for a given choice of K and ν, the function u determined by (10) solves $\Delta u = 4\,u^2$ in D. This is analogous to the beginning of the proof of Proposition 1. Recall the notation $K_\varepsilon = \{y \in \partial D, \text{dist}(y, K) < \varepsilon\}$. We can choose a sequence of functions $g_n \in C_0(\partial D \backslash K)$ such that the measures $g_n(y)\theta(dy)$ converge vaguely to $\nu(dy)$ as $n \to \infty$. Set $h_n = n\,1_{K_\varepsilon} + g_n$. Then, for every n the function

$$u_n^\varepsilon(x) = \mathbb{N}_x(1 - \exp - \langle \mathcal{Z}^D, h_n \rangle), \quad x \in D$$

solves $\Delta u = 4\,u^2$ in D.
As in the proof of Proposition 1, we have $\lim_{n\to\infty}\langle \mathcal{Z}^D, h_n \rangle = \infty$ a.e. on $\{\mathcal{E}^D \cap K_\varepsilon \neq \emptyset\}$. On the other hand, on $\{\mathcal{E}^D \cap K_\varepsilon = \emptyset\}$, the function $z_D(y)$ has compact support in $\partial D \backslash K$, and the vague convergence of $g_n(y)\theta(dy)$ towards $\nu(dy)$ gives

$$\lim_{n\to\infty} \langle \mathcal{Z}^D, h_n \rangle = \lim_{n\to\infty} \langle \mathcal{Z}^D, g_n \rangle = \lim_{n\to\infty} \int \theta(dy)\, z_D(y) g_n(y) = \int \nu(dy)\, z_D(y).$$

By combining the previous observations, we get

$$\lim_{n\to\infty} u_n^\varepsilon(x)$$

$$= \mathbb{N}_x(\mathcal{E}^D \cap K_\varepsilon \neq \emptyset) + \mathbb{N}_x\left(1_{\{\mathcal{E}^D \cap K_\varepsilon = \emptyset\}}\left(1 - \exp-\int \nu(dy)\, z_D(y)\right)\right) =: u^\varepsilon(x).$$

By Proposition V.9 (iii), this implies that u^ε is a solution, and so is $u = \lim \downarrow u^\varepsilon$.

Step 2. We now construct the pair (K, ν) for a given solution. From now on until the end of the proof, we fix a nonnegative solution u. We have to check that u can be written in the form (10) and that the pair (K, ν) is determined from u by the formulas of Theorem 5.

We choose a sequence (r_n) converging to 1 so that the conclusion of Lemma 4 holds. Then, for $p, q \in \mathbb{T} = \mathbb{R}/2\pi\mathbb{Z}$, we set

$$a_n(p, q) = \int_{(p,q)} u(r_n\, e^{i\beta})\, d\beta.$$

(We use the obvious convention for intervals in $\mathbb{T}$: If a, respectively b, is the representative of p, respectively q, in $[0, 2\pi)$, we take $(p, q) = (a, b)$ if $a \leq b$, $(p, q) = (a, b + 2\pi)$ if $a > b$). Replacing (r_n) by a subsequence if necessary, we may assume that, for every $p, q \in \mathbb{T}_1 := \mathbb{T} \cap 2\pi\mathbb{Q}$,

$$\lim_{n\to\infty} a_n(p, q) = a(p, q) \in \mathbb{R}_+ \cup \{+\infty\}.$$

Note that $a(p, r) = a(p, q) + a(q, r)$ if $p, q, r \in \mathbb{T}_1$ and $q \in (p, r)$. We set

$$K = \{y \in \mathbb{T},\ a(p, q) = \infty \text{ whenever } p, q \in \mathbb{T}_1 \text{ and } y \in (p, q)\}.$$

Then K is a compact subset of $\mathbb{T}$, which is identified to ∂D.

We also set $O = \mathbb{T}\backslash K$ and define a finite measure ν_n on O by

$$\nu_n(d\beta) = 1_O(\beta)\, u(r_n e^{i\beta})\, d\beta.$$

From the definition of K, we see that for every compact subset H of O,

$$\sup_n \nu_n(H) < \infty.$$

Hence, by extracting again a subsequence, we may assume that the sequence (ν_n) converges vaguely in the space of Radon measures on O. Through the identification $\partial D = \mathbb{T}$, the limiting measure ν is a Radon measure on $\partial D\backslash K$.

Step 3. We now prove that formula (10) holds for the given solution u and the pair (K, ν) introduced in Step 2. To simplify notation, we will write $z_D(r, \beta)$ instead of $z_D(r, e^{i\beta})$ and $z_D(\beta)$ instead of $z_D(e^{i\beta})$. By Corollary V.8, we have for $x \in D_{r_n}$

$$u(x) = \mathbb{N}_x\left(1 - \exp -\langle \mathcal{Z}^{D_{r_n}}, u\rangle\right) = \mathbb{N}_x\left(1 - \exp -r_n \int_{\mathbb{T}} d\beta\, u(r_n e^{i\beta})\, z_D(r_n, \beta)\right).$$

Lemma 6. *For every $x \in D$, we have*

$$\lim_{n \to \infty} \int_{\mathbb{T}} d\beta\, u(r_n e^{i\beta}))\, z_D(r_n, \beta) = +\infty, \quad \mathbb{N}_x \text{ a.e. on } \{\mathcal{E}^D \cap K \neq \emptyset\}, \quad (11)$$

and

$$\lim_{n \to \infty} \int_{\mathbb{T}} d\beta\, u(r_n e^{i\beta}))\, z_D(r_n, \beta) = \int_{\mathbb{T}} \nu(d\beta)\, z_D(\beta), \quad \mathbb{N}_x \text{ a.e. on } \{\mathcal{E}^D \cap K = \emptyset\}. \quad (12)$$

Observe that formula (10) immediately follows from Lemma 6 by passing to the limit $n \to \infty$ in the preceding formula for $u(x)$. To complete Step 3, it remains to prove Lemma 6.

Proof of Lemma 6. We first prove (12). For $\delta > 0$, denote by U_δ the open tubular neighborhood of radius δ of K in $\mathbb{R}^2$. Also introduce the random set

$$\mathcal{R}^D = \{y = W_s(t), 0 \leq s \leq \sigma, 0 \leq t \leq \zeta_s \wedge \tau(W_s)\}.$$

Note that $\mathcal{E}^D = \mathcal{R}^D \cap \partial D$.

Since $\mathcal{R}^D$ is compact, on the set $\{\mathcal{E}^D \cap K = \emptyset\}$ we may find $\delta = \delta(\omega) > 0$ so small that $\mathcal{R}^D \cap U_\delta = \emptyset$. It follows that $z_D(r, y) = 0$ for every $y \in \partial D$ and $r \in (0, 1]$ such that $ry \in U_\delta$. Choosing $\varepsilon = \delta/2$, we see that $z_D(r, y) = 0$ for every $y \in K_\varepsilon$ and $r \in (1 - \varepsilon, 1]$.
From the definition of K we have

$$\sup_n \int_{\mathbb{T} \setminus K_\varepsilon} d\beta\, u(r_n e^{i\beta}) < \infty.$$

Then Lemma 4 implies

$$\lim_{n \to \infty} \int_{\mathbb{T}} d\beta\, u(r_n e^{i\beta})\, |z_D(r_n, \beta) - z_D(\beta)| = 0,$$

$\mathbb{N}_x$ a.e. on the set $\{\mathcal{E}^D \cap K = \emptyset\}$. On the other hand, on the same event we have for n large

$$\int_{\mathbb{T}} d\beta\, u(r_n e^{i\beta})\, z_D(\beta) = \int_{\mathbb{T}} \nu_n(d\beta)\, z_D(\beta)$$

which converges to $\int_{\mathbb{T}} \nu(d\beta)\, z_D(\beta)$ by the vague convergence of ν_n towards ν. This completes the proof of (12).

Unfortunately, the proof of (11) is more involved. We first introduce the stopping time

$$T = \inf\{s \geq 0,\, \zeta_s = \tau(W_s),\, \hat{W}_s \in K\},$$

in such a way that $\{\mathcal{E}^D \cap K \neq \emptyset\} = \{T < \infty\}$, a.e., and $\hat{W}_T \in K$ a.e. on the event $\{T < \infty\}$. We will prove that

$$z_D(\hat{W}_T) > 0, \quad \mathbb{N}_x \text{ a.e. on } \{T < \infty\}. \tag{13}$$

Our claim (11) easily follows from (13): By Lemma 4 and the continuity of $z_D(\beta)$, we may find $\varepsilon > 0$ such that $z_D(r_n, \beta) \geq \frac{1}{2} z_D(\hat{W}_T)$ for every n sufficiently large and $|\beta - \hat{W}_T| < \varepsilon$. Thus, for n large,

$$\int_{\mathbb{T}} d\beta\, u(r_n e^{i\beta})\, z_D(r_n, \beta) \geq \frac{1}{2} z_D(\hat{W}_T) \int_{|\beta - \hat{W}_T| < \varepsilon} d\beta\, u(r_n e^{i\beta})$$

which tends to ∞ by the property $\hat{W}_T \in K$ and the definition of K.

In order to prove (13), we apply the strong Markov property to the Brownian snake at time T and use Lemma V.5. We need to control the behavior of the path W_T near its endpoint. To this end, we will use both Lemma VI.10 and another technical result showing that the path W_T cannot be "too close" to the boundary immediately before $\zeta_T = \tau(W_T)$.

Lemma 7. *Let* $x \in D$. *We can choose* $\alpha > 0$ *so that* $\mathbb{N}_x$ *a.e. for every* $s \in (0, \sigma)$ *such that* $\tau(W_s) \geq \zeta_s$,

$$\liminf_{m \to \infty} \frac{1}{m} \mathrm{Card}\{p \leq m;\, W_s(t) \in D_{1-\alpha 2^{-p}},$$

$$\forall t \in [(\zeta_s - 2^{-2p})^+, (\zeta_s - 2^{-2p-1})^+]\} > \frac{1}{2}.$$

We postpone the proof of Lemma 7 to the end of this section. As a consequence of Lemma VI.10 and Lemma 7, the stopped path W_T satisfies the following two properties $\mathbb{N}_x$ a.e. on $\{T < \infty\}$:

(a) $\tau(W_T) = \zeta_T$ and $\hat{W}_T \in K$.

(b) There exist positive constants α and A such that the property

$$\{W_T(\zeta_T - t), 2^{-2p-1} \le t \le 2^{-2p}\} \subset D_{1-\alpha 2^{-p}} \cap B(\hat{W}_T, A\, 2^{-p})$$

holds for infinitely many $p \in \mathbb{N}$.

We now fix a stopped path $w \in \mathcal{W}_x$ such that (a) and (b) hold when W_T and ζ_T are replaced by w and ζ_w respectively . Write $\mathbb{P}^*_w$ for the law of the Brownian snake started at w and stopped when its lifetime process vanishes. Lemma V.5 allows us to define $z_D(y)$, $y \in D$ under $\mathbb{P}^*_w$ via the formula $z_D(y) = \sum_{i \in I} z_D(y)(W^i)$. From Lemma V.5, we have then

$$\mathbb{P}^*_w(z_D(\hat{w}) > 0) = 1 - \exp -2 \int_0^{\zeta_w} dt\, \mathbb{N}_{w(t)}(z_D(\hat{w}) > 0). \tag{14}$$

However, for $a \in D$,

$$\mathbb{N}_a(z_D(\hat{w}) > 0) \ge \frac{\left(\mathbb{N}_a(z_D(\hat{w}))\right)^2}{\mathbb{N}_a(z_D(\hat{w})^2)} = \frac{P_D(a, \hat{w})^2}{4 \int_D dy\, G_D(a, y)\, P_D(y, \hat{w})^2}, \tag{15}$$

using the Cauchy-Schwarz inequality and Proposition 3.

On one hand, easy calculations using the explicit formulas (1) give the existence of a constant C such that, for every $a \in D$ and $z \in \partial D$,

$$\int_D dy\, G_D(a, y)\, P_D(y, z)^2 \le C. \tag{16}$$

On the other hand, if $a \in D_{1-\alpha 2^{-p}} \cap B(\hat{w}, A\, 2^{-p})$, we have

$$P_D(a, \hat{w}) = \frac{1}{2\pi} \frac{1 - |a|^2}{|\hat{w} - a|^2} \ge \frac{\alpha}{2\pi A^2}\, 2^p.$$

By substituting these estimates in (15), we get that for every integer p such that $2^{-2p} \le \zeta_w$ and

$$\{w(\zeta_w - t),\, 2^{-2p-1} \le t \le 2^{-2p}\} \subset D_{1-\alpha 2^{-p}} \cap B(\hat{w}, A\, 2^{-p}),$$

we have

$$\int_{\zeta_w - 2^{-2p}}^{\zeta_w - 2^{-2p-1}} dt\, \mathbb{N}_{w(t)}(z_D(\hat{w}) > 0) \ge \frac{1}{8C}\left(\frac{\alpha}{2\pi A^2}\right)^2.$$

By (b) this lower bound holds for infinitely many values of p, and we conclude from (14) that $\mathbb{P}_w^*(z_D(\hat{w}) > 0) = 1$. Our claim (13) now follows by applying the strong Markov property at time T.

Step 4. It remains to verify that K and ν are determined from u by formulas (8) and (9) of Theorem 5 (this will in particular give the uniqueness of the pair (K, ν), which is not clear from the previous construction). We rely on formula (10). First, if $y \in K$, we have for every $x \in D$,

$$u(x) \geq \mathbb{N}_x(y \in \mathcal{E}^D) \geq \mathbb{N}_x(z_D(y) > 0)$$
$$\geq \frac{P_D(x,y)^2}{4 \int_D da\, G_D(x,a) P_D(a,y)^2} \geq \frac{P_D(x,y)^2}{4C},$$

by the arguments we have just used in Step 3. The fact that K is contained in the set in the right side of (8) immediately follows from this estimate. The converse inclusion is clear from our definition of K.

Let us prove (9). Let $g \in C_{b+}(\partial D)$ be such that $\operatorname{supp} g$ is contained in $\partial D \backslash K$. By Proposition 1,

$$\lim_{r \uparrow 1, r < 1} \int \theta(dy)\, g(y)\, \mathbb{N}_{ry}(\mathcal{E}^D \cap K \neq \emptyset) = 0. \tag{17}$$

As a consequence of (10) and (17), we have

$$\lim_{r \uparrow 1, r < 1} \left| \int \theta(dz) g(z) u(rz) - \int \theta(dz) g(z)\, \mathbb{N}_{rz}\left(1 - \exp - \int \nu(dy) z_D(y)\right) \right| = 0.$$

Let $\varepsilon > 0$ be such that $g = 0$ on K_ε. Denote by ν' the restriction of ν to $K_{\varepsilon/2}$, so that ν' is a finite measure and $\langle \nu', g \rangle = \langle \nu, g \rangle$. Furthermore,

$$\left| \int \theta(dz) g(z)\, \mathbb{N}_{rz}\left(1 - \exp - \int \nu'(dy) z_D(y)\right) \right.$$
$$\left. - \int \theta(dz) g(z)\, \mathbb{N}_{rz}\left(1 - \exp - \int \nu(dy) z_D(y)\right) \right|$$
$$\leq \int \theta(dz) g(z) \sup_{z \in \partial D \backslash K_\varepsilon} \mathbb{N}_{rz}(\mathcal{E}^D \cap K_{\varepsilon/2} \neq \emptyset),$$

and Proposition 1 again shows that the latter quantity goes to 0 as $r \uparrow 1$. In view of these considerations, the proof of (9) reduces to checking that

$$\langle \nu', g \rangle = \lim_{r \uparrow 1, r < 1} \int \theta(dz) g(z)\, \mathbb{N}_{rz}\left(1 - \exp - \int \nu'(dy) z_D(y)\right). \tag{18}$$

First note that

$$\int \theta(dz)g(z)\,\mathbb{N}_{rz}\Big(\int \nu'(dy)z_D(y)\Big) = \int \theta(dz)g(z)\int \nu'(dy)P_D(rz,y)$$

$$= \int \theta(dz)g(z)\int \nu'(dy)P_D(ry,z)$$

$$= \int \nu'(dy)\Pi_{ry}\big(g(\xi_\tau)\big)$$

tends to $\langle \nu',g\rangle$ as $r \uparrow 1$. Thus,

$$\int \theta(dz)g(z)\,\mathbb{N}_{rz}\Big(1 - \exp - \int \nu'(dy)z_D(y)\Big)$$

$$\leq \int \theta(dz)g(z)\,\mathbb{N}_{rz}\Big(\int \nu'(dy)z_D(y)\Big) = \langle \nu',g\rangle.$$

On the other hand, we have for any $\eta > 0$,

$$\int \theta(dz)g(z)\,\mathbb{N}_{rz}\Big(\int \nu'(dy)z_D(y)1_{\{\int \nu'(dy)z_D(y)>\eta\}}\Big)$$

$$\leq \eta^{-1}\int \theta(dz)g(z)\,\mathbb{N}_{rz}\Big(\big(\int \nu'(dy)z_D(y)\big)^2\Big)$$

$$\leq \langle \nu',1\rangle\eta^{-1}\int \theta(dz)g(z)\,\mathbb{N}_{rz}\Big(\int \nu'(dy)\,z_D(y)^2\Big)$$

$$= \langle \nu',1\rangle\eta^{-1}\int \theta(dz)g(z)\int \nu'(dy)\int_D da\,G_D(rz,a)\,P_D(a,y)^2.$$

The last quantity tends to 0 as $r \uparrow 1$ by dominated convergence, using (16) and the (easy) fact that if $z \neq y$,

$$\lim_{r\uparrow 1, r<1}\int_D da\,G_D(rz,a)\,P_D(a,y)^2 = 0.$$

Then notice that, for every $\gamma > 0$, we can choose $\eta > 0$ small enough so that

$$\int \theta(dz)g(z)\,\mathbb{N}_{rz}\Big(1 - \exp - \int \nu'(dy)z_D(y)\Big)$$

$$\geq (1-\gamma)\int \theta(dz)g(z)\,\mathbb{N}_{rz}\Big(\int \nu'(dy)z_D(y)1_{\{\int \nu'(dy)z_D(y)\leq\eta\}}\Big)$$

and so it follows from the previous estimates that

$$\liminf_{r\uparrow 1, r<1}\int \theta(dz)g(z)\,\mathbb{N}_{rz}\Big(1 - \exp - \int \nu'(dy)z_D(y)\Big) \geq (1-\gamma)\langle \nu',g\rangle.$$

Since γ was arbitrary, this completes the proof of (18) and of Theorem 5. $\square$

Proof of Lemma 7. This is very similar to the proof of Lemma VI.10. We use again the stopping times σ_i^n defined in this proof and start with the following simple observation. Let $s \in (0, \sigma)$ and $n \geq 3$ such that $2.2^{-2n} < \zeta_s$. Let $k \geq 2$ be such that $k2^{-2n} \leq s < (k+1)2^{-2n}$. There is a random integer j such that

$$\inf\{s' > s, \zeta_{s'} = (k-1)2^{-2n}\} = \sigma_j^n.$$

By the snake property, $W_{\sigma_j^n}$ is the restriction of W_s to the interval $[0, (k-1)2^{-n}]$. It easily follows that, for every $p \in \{1, \ldots, n-2\}$,

$$\{W_s(t), t \in [(\zeta_s - 2^{-2p})^+, (\zeta_s - 2^{-2p-1})^+]\}$$
$$\subset \{W_{\sigma_j^n}(t), t \in [(\zeta_{\sigma_j^n} - 2^{-2p})^+, (\zeta_{\sigma_j^n} - \frac{3}{2}2^{-2p-2})^+]\}.$$

As a consequence of this remark, the statement of the lemma will follow if we can prove that for a suitable value of $\alpha > 0$ we have N_x a.e., for all n sufficiently large and all j such that $\sigma_j^n < \infty$ and $\tau(W_{\sigma_j^n}) = \infty$,

$$\text{Card}\{p \leq n \,;\, W_{\sigma_j^n}(t) \in D_{1-\alpha 2^{-p}}, \forall t \in [(\zeta_{\sigma_j^n} - 2^{-2p})^+, (\zeta_{\sigma_j^n} - \frac{3}{2}2^{-2p-2})^+]\} > \frac{n}{2}. \tag{19}$$

For every stopped path $w \in \mathcal{W}_x$, set

$$G_n^\alpha(w) = \frac{1}{n}\text{Card}\{p \leq n; w(t) \in D_{1-\alpha 2^{-p}}, \forall t \in [(\zeta_w - 2^{-2p})^+, (\zeta_w - \frac{3}{2}2^{-2p-2})^+]\}.$$

By the same arguments as in the proof of Lemma VI.10, our claim (19) will follow if we can get a good estimate on $\Pi_x(\tau > t, G_n^\alpha(\xi_{[0,t]}) \leq n/2)$. Precisely, it is enough to show that, for every $\lambda > 0$, we can choose $\alpha > 0$ sufficiently small so that, for every $n \geq 2$ and every $t > 0$,

$$\Pi_x(\tau > t, G_n^\alpha(\xi_{[0,t]}) \leq n/2) \leq e^{-\lambda n} \tag{20}$$

(compare with Lemma VI.11). Set $m = [n/2]$ and observe that

$$\Pi_x(\tau > t, G_n^\alpha(\xi_{[0,t]}) \leq n/2) \leq \sum_{1 \leq k_1 < k_2 < \ldots < k_m \leq n} \Pi_x(\mathcal{U}_{k_1} \cap \ldots \cap \mathcal{U}_{k_m}) \tag{21}$$

where for every $p \in \{1, \ldots, n\}$, $\mathcal{U}_p$ denotes the event

$$\mathcal{U}_p = \{\xi_r \in D, \forall r \in [(t - 2^{-2p})^+, (t - 2^{-2p-2})^+]\}$$
$$\cap \{\exists r \in [(t - 2^{-2p})^+, (t - \frac{3}{2}2^{-2p-2})^+] : \xi_r \notin D_{1-\alpha 2^{-p}}\}.$$

By applying the strong Markov property at

$$\inf\{r \geq t - 2^{-2k_m}, \, \xi_r \notin D_{1-\alpha 2^{-k_m}}\},$$

we get

$$\Pi_x(\mathcal{U}_{k_1} \cap \ldots \cap \mathcal{U}_{k_m})$$

$$\leq \Pi_x(\mathcal{U}_{k_1} \cap \ldots \cap \mathcal{U}_{k_{m-1}}) \times \sup_{a \in D \setminus D_{1-\alpha 2^{-k_m}}} \Pi_a(\xi_r \in D, \, \forall r \in [0, 2^{-2k_m-3}])$$

$$\leq c(\alpha) \, \Pi_x(\mathcal{U}_{k_1} \cap \ldots \cap \mathcal{U}_{k_{m-1}})$$

with a constant $c(\alpha)$ depending only on α and such that $c(\alpha) \to 0$ as $\alpha \to 0$. By iterating this argument, we arrive at the estimate

$$\Pi_x(\mathcal{U}_{k_1} \cap \ldots \cap \mathcal{U}_{k_m}) \leq c(\alpha)^m = c(\alpha)^{[n/2]}.$$

By susbtituting this estimate in (21), and choosing a suitable value of α, we get the bound (20). This completes the proof of Lemma 7.   □

4   Further developments

A number of recent papers have studied extensions of Theorem 5 to higher dimensions and more general equations. The purpose of these papers is usually to define the trace of a nonnegative solution (possibly belonging to a special class) and then to study the properties of the map that associates with a solution its trace. In this section, we briefly survey these recent developments. The word solution always means nonnegative solution.

A solution u of $\Delta u = 4u^2$ in a domain D is called moderate if it is bounded above by a function harmonic in D. In the setting of Theorem 5, this corresponds to the case $K = \emptyset$ (which implies that ν is finite), and the minimal harmonic majorant of u is then the function $h(x) = \int \nu(dy) P_D(x, y)$. For a general smooth domain in dimension d, one can show [L9] that moderate solutions are in one-to-one correspondence with finite measures ν on ∂D that do not charge boundary polar sets (this result had been conjectured by Dynkin [D7]). This correspondence has been extended by Dynkin and Kuznetsov [DK3] to the equation $\Delta u = u^p$, $1 < p \leq 2$ (in fact to $Lu = u^p$ for a general elliptic operator L). In this more general setting, the notion of boundary polar sets is defined in terms of the superprocess with branching mechanism $\psi(u) = u^p$, and an analytic characterization analogous to Theorem 2 holds [DK2].

The analytic part of Theorem 5 has been extended by Marcus and Véron [MV2] to the equation $\Delta u = u^p$ $(p > 1)$ in the unit ball of $\mathbb{R}^d$, provided that

$d < \frac{p+1}{p-1}$. In this case, the so-called subcritical case, all assertions of Theorem 5 remain true, except of course the probabilistic formula (10).

The supercritical case $d \geq \frac{p+1}{p-1}$ ($d \geq 3$ when $p = 2$) is more difficult and more interesting. Marcus and Véron [MV3] and Dynkin and Kuznetsov [DK4] (see also [DK5] for extensions to a general domain on a Riemannian manifold) have shown that it is still possible to define the trace as a pair (K, ν) as in Theorem 5. In fact, formulas (8) and (9) can be used for this purpose, with obvious modifications when D is a general smooth domain. There are however two essential differences with the subcritical case:

(i) Not all pairs (K, ν) are admissible. For instance, if $p = 2$, the pair $(K, 0)$ is not admissible when K is boundary polar. Dynkin and Kuznetsov [DK4] (in the case $1 < p \leq 2$) and Marcus and Véron [MV1] have independently described all possible traces. When $1 < p \leq 2$, a probabilistic formula analogous to (10) holds for the maximal solution associated with a given trace.

(ii) Infinitely many solutions may have the same trace. Here is an example, adapted from [L10], in the case when $p = 2$ and D is the unit ball in $\mathbb{R}^d$, $d \geq 3$. Let (y_n) be a dense sequence in ∂D and, for every n, let $(r_n^p, p = 1, 2, \ldots)$ be a decreasing sequence of positive numbers. Recall the notation $N_r(y) = \{z \in \partial D, |z - y| < r\}$ and, for every $p \geq 1$, set

$$H_p = \bigcup_{n=1}^{\infty} N_{r_n^p}(y_n),$$

$$u_p(x) = \mathbb{N}_x(\mathcal{E}^D \cap H_p \neq \emptyset), \qquad x \in D.$$

Then it is easy to see that, for every $p \geq 1$, u_p is a solution with trace $(\partial D, 0)$. On the other hand, the fact that singletons are boundary polar implies that $u_p \downarrow 0$ as $p \uparrow \infty$, provided that the sequences $(r_n^p, p = 1, 2, \ldots)$ decrease sufficiently fast. Therefore infinitely many of the functions u_p must be different.

In view of this nonuniqueness problem, Dynkin and Kuznetsov [Ku], [DK6] have proposed to use a finer definition of the trace, where the set K is no longer closed with respect to the Euclidean topology. This finer definition leads to a one-to-one correspondence between solutions and possible traces, provided that one considers only σ-moderate solutions: A solution is σ-moderate if and only if it is the limit of an increasing sequence of moderate solutions. An intriguing open problem is whether there exist solutions that are not σ-moderate.

We refer to the survey [DK8] for a more detailed account of the recent results and open problems in this area.

Chapter VIII
Lévy Processes and the Genealogy of General Continuous-state Branching Processes

The Brownian snake construction of quadratic superprocesses relies on the fact that the genealogical structure of the Feller diffusion can be coded by reflected Brownian motion. Our goal in this chapter is to explain a similar coding for the genealogy of continuous-state branching processes with a general branching mechanism ψ. The role of reflected Brownian motion will be played by a certain functional of a Lévy process with no negative jumps and Laplace exponent ψ. We first explain the key underlying ideas in a discrete setting.

1 The discrete setting

We consider an offspring distribution μ, that is a probability measure on $\mathbb{N} = \{0, 1, 2, \ldots\}$. We assume that $\mu(1) < 1$ and that μ is critical or subcritical: $\sum k\mu(k) \leq 1$.

The law of the Galton-Watson tree with offspring distribution μ, in short the μ-Galton-Watson tree, can then be realized as a probability distribution on the set of all finite trees. Here, a (finite) tree is a finite subset $\mathcal{T}$ of $\bigcup_{n=0}^{\infty}(\mathbb{N}^*)^n$ (where $\mathbb{N}^* = \{1, 2, \ldots\}$ and $(\mathbb{N}^*)^0 = \{\phi\}$) which satisfies the obvious properties:

(i) $\phi \in \mathcal{T}$ (ϕ is the root of $\mathcal{T}$).

(ii) If $(u_1, \ldots, u_n) \in \mathcal{T}$ with $n \geq 1$, then $(u_1, \ldots, u_{n-1}) \in \mathcal{T}$.

(iii) If $u = (u_1, \ldots, u_n) \in \mathcal{T}$, there exists an integer $k_u(\mathcal{T}) \geq 0$ such that $(u_1, \ldots, u_n, k) \in \mathcal{T}$ if and only if $k \leq k_u(\mathcal{T})$.

If $u \in (\mathbb{N}^*)^n$, the generation of u is $|u| = n$.

Consider then a sequence $\mathcal{T}_0, \mathcal{T}_1, \ldots, \mathcal{T}_k, \ldots$ of independent μ-Galton-Watson trees. We can code this sequence by the following procedure. We consider a "particle" that visits the vertices of $\mathcal{T}_0, \ldots, \mathcal{T}_k, \ldots$ according to the following rules:

- The particle starts at time $n = 0$ from the root of $\mathcal{T}_0$ then visits all other vertices of $\mathcal{T}_0$, then the vertices of $\mathcal{T}_1$, and so on.

- For each tree, the particle visits its vertices successively in lexicographical order.

Denote by H_n the generation of the vertex that is visited at time n. It is easy to see that the function $n \to H_n$ provides a coding of the sequence of trees. (See Fig. I.6 for an example.) We then want to have a better probabilistic understanding of this coding.

Proposition 1. *There exists a random walk $V = (V_n, n \geq 0)$ on $\mathbb{Z}$ with jump distribution $\nu(k) = \mu(k+1)$, $k = -1, 0, 1, \ldots$ such that for every $n \geq 0$*

$$H_n = \mathrm{Card}\{j \in \{0, 1, \ldots, n-1\}, V_j = \inf_{j \leq k \leq n} V_k\}.$$

Proposition 1 is elementary. Let us outline the ingredients of the proof. For every $j \in \{1, \ldots, H_n\}$ let $\rho_j(n)$ be the number of "younger brothers" of the ancestor of the individual visited at time n in the j-th generation. More precisely, if $u(n) = (u_1(n), \ldots, u_{H_n}(n))$ is the vertex visited at time n, and $\mathcal{T}_{\ell(n)}$ is the tree to which it belongs, we set for every $j = 1, \ldots, H_n$,

$$\rho_j(n) = \mathrm{Card}\{k > u_j(n); (u_1(n), \ldots, u_{j-1}(n), k) \in \mathcal{T}_{\ell(n)}\}.$$

Set $\rho(n) = (\rho_1(n), \ldots, \rho_{H_n}(n))$. When $H_n = 0$, $\rho(n) = \emptyset$ is the empty sequence. Then it is easy to see that $\rho(n)$ is a Markov chain in the set of finite sequences of nonnegative integers, with transition kernel given as follows. For $k \geq 0$,

$$P[\rho(n+1) = (\alpha_1, \ldots, \alpha_p, k) \mid \rho(n) = (\alpha_1, \ldots, \alpha_p)] = \mu(k+1)$$

and, if $q = \sup\{j, \alpha_j > 0\}$ ($\sup \emptyset = 0$),

$$P[\rho(n+1) = (\alpha_1, \ldots, \alpha_q - 1) \mid \rho(n) = (\alpha_1, \ldots, \alpha_p)] = \mu(0),$$

with the convention that $(\alpha_1, \ldots, \alpha_q - 1) = \emptyset$ if $q = 0$. The first formula corresponds to the case when the individual (vertex) visited at time n has $k + 1$ children: Then the individual visited at time $n+1$ will be the first of these children. The second formula corresponds to the case when the individual visited at time n has no child: Then the next visited individual is the "first available brother", namely $(u_1(n), \ldots, u_{q-1}(n), u_q(n) + 1)$ in the previous notation (if $q = 0$ it is the root of the next tree). The Markov property for $(\rho(n), n \geq 0)$ comes from the fact that, at the time when we visit an individual, the past gives us no information on the number of its children (this is so because of the lexicographical order of visits).

The random walk $(V_n, n \geq 0)$ can be defined in terms of $(\rho(n), n \geq 0)$ by the formula

$$V_n = \sum_{j=1}^{H_n} \rho_j(n) - \ell(n) = \sum_{j=1}^{H_n} \rho_j(n) - \mathrm{Card}\{k \in \{1, \ldots, n\}, \rho(k) = \emptyset\} .$$

The fact that it has the desired distribution easily follows from the formulas for the transition kernel of $\rho(n)$. Also observe that

$$\langle \rho(n), 1 \rangle := \sum_{j=1}^{H_n} \rho_j(n) = V_n - \inf_{0 \leq j \leq n} V_j$$

is the reflected random walk.

Finally, the explicit formula for H in terms of V is easy to derive. Note that the condition

$$V_j = \inf_{j \leq k \leq n} V_k$$

holds iff $n < \inf\{k > j, V_k < V_j\}$. But the latter infimum is the first time of visit of an individual that is not a descendant of $u(j)$ (as long as we are visiting descendants of $u(j)$, the "total number of younger brothers" $\langle \rho(k), 1 \rangle$ is at least as large as $\langle \rho(j), 1 \rangle$). Hence the condition $n < \inf\{k > j, V_k < V_j\}$ holds iff $u(n)$ is a descendant of $u(j)$, or equivalently $u(j)$ is an ancestor of $u(n)$. Therefore

$$\mathrm{Card}\{j \in \{0, 1, \ldots, n-1\}, V_j = \inf_{j \leq k \leq n} V_k\}$$

is the number of ancestors of $u(n)$, and is thus equal to H_n.

Our main goal in the following sections will be to study a continuous analogue of the previous coding. The role of the random walk $(V_n, n \geq 0)$ will be played by a Lévy process with no negative jumps.

Exercise. Verify that an invariant measure for $(\rho(n), n \geq 0)$ is

$$M\big((\alpha_1, \ldots, \alpha_p)\big) = \bar{\mu}(\alpha_1) \cdots \bar{\mu}(\alpha_p) \quad \text{where } \bar{\mu}(j) = \mu\big((j, \infty)\big) .$$

2   Lévy processes

In this section we introduce the class of Lévy processes that will be relevant to our purposes and we record some of their properties.

We start from a function ψ of the type considered in Chapter II:

$$\psi(\lambda) = \alpha\lambda + \beta\lambda^2 + \int_{(0,\infty)} \pi(dr)(e^{-\lambda r} - 1 + \lambda r)$$

where $\alpha \geq 0$, $\beta \geq 0$ and π is a σ-finite measure on $(0, \infty)$ such that $\int \pi(dr)(r \wedge r^2) < \infty$ (cf. Theorem II.1).

Then there exists a Lévy process (real-valued process with stationary indepen-dent increments) $Y = (Y_t, t \geq 0)$ started at $Y_0 = 0$, whose Laplace exponent is ψ, in the sense that for every $t \geq 0$, $\lambda \geq 0$:

$$E[e^{-\lambda Y_t}] = e^{t\psi(\lambda)} .$$

The measure π is the Lévy measure of Y, β corresponds to its Brownian part, and $-\alpha$ to a drift coefficient (after compensation of the jumps). Since π is supported on $(0, \infty)$, Y has no negative jumps. In fact, under our assumptions, Y can be the most general Lévy process without negative jumps that does not drift to $+\infty$ (i.e. we cannot have $Y_t \to \infty$ as $t \to \infty$, a.s.). This corresponds to the fact that we consider only critical or subcritical branching.

The point 0 is always regular for $(-\infty, 0)$, with respect to Y, meaning that

$$P\big(\inf\{t > 0, Y_t < 0\} = 0\big) = 1 .$$

It is not always true that 0 is regular for $(0, \infty)$, but this holds if

$$\beta > 0 , \text{ or } \beta = 0 \text{ and } \int_0^1 r\pi(dr) = \infty . \tag{1}$$

From now on we will assume that (1) holds. This is equivalent to the property that the paths of Y are a.s. of infinite variation. A parallel theory can be developed in the finite variation case, but the cases of interest in relation with superprocesses (the stable case where $\pi(dr) = cr^{-2-\alpha}dr, 0 < \alpha < 1$) do satisfy (1).

Consider the maximum and minimum processes of Y:

$$S_t = \sup_{s \leq t} Y_s , \quad I_t = \inf_{s \leq t} Y_s .$$

Both $S-Y$ and $Y-I$ are Markov processes in $\mathbb{R}_+$ (this is true indeed for any Lévy process). From the previous remarks on the regularity of 0, it immediately follows that 0 is a regular point (for itself) with respect to both $S-Y$ and $Y-I$. We can therefore consider the (Markov) local time of both $S-Y$ and $Y-I$ at level 0.

It is easy to see that the process $-I$ provides a local time at 0 for $Y-I$. We will denote by N the associated excursion measure. By abuse of notation, we still denote by Y the canonical process under N. Under N, Y takes nonnegative values and $Y_t > 0$ if and only if $0 < t < \sigma$, where σ denotes the duration of the excursion.

We denote by $L = (L_t, t \geq 0)$ the local time at 0 of $S-Y$. Here we need to specify the normalization of L. This can be done by the following approximation:

$$L_t = \lim_{\varepsilon \downarrow 0} \frac{1}{\varepsilon} \int_0^t 1_{\{S_s - Y_s < \varepsilon\}} ds \, , \tag{2}$$

in probability. If $L^{-1}(t) = \inf\{s, L_s > t\}$ denotes the right-continuous inverse of L, formula (2) follows from the slightly more precise result

$$\lim_{\varepsilon \to 0} E\left[\left(\frac{1}{\varepsilon} \int_0^{L^{-1}(t)} 1_{\{S_s - Y_s < \varepsilon\}} ds - (t \wedge L_\infty)\right)^2\right] = 0 \tag{2$'$}$$

which can be derived from excursion theory for $S-Y$ (after choosing the proper normalization for L).

The process $(S_{L^{-1}(t)}, t \geq 0)$ is a subordinator (that is, a Lévy process with nondecreasing paths) and a famous formula of fluctuation theory gives its Laplace transform

$$E(\exp -\lambda S_{L^{-1}(t)}) = \exp\left(-t\frac{\psi(\lambda)}{\lambda}\right) . \tag{3}$$

Note that

$$\frac{\psi(\lambda)}{\lambda} = \alpha + \beta\lambda + \int_0^\infty dr\, \pi((r,\infty))(1 - e^{-\lambda r})$$

so that the subordinator $(S_{L^{-1}(t)}, t \geq 0)$ has Lévy measure $\pi((r,\infty))dr$, drift β and is killed at rate α. In particular for every $s \geq 0$, if m denotes the Lebesgue measure on $\mathbb{R}_+$, we have a.s.

$$m(\{S_{L^{-1}(r)}; 0 \leq r \leq s, L^{-1}(r) < \infty\}) = \beta(s \wedge L_\infty)$$

from which it easily follows that

$$m(\{S_r, 0 \leq r \leq t\}) = \beta L_t \, . \tag{4}$$

Note that when $\beta > 0$ this formula yields an explicit expression for L_t.

3   The height process

Recall the formula of Proposition 1 above. If we formally try to extend this formula to our continuous setting, replacing the random walk V by the Lévy process Y, we are lead to define H_t as the Lebesgue measure of the set $\{s \leq t, Y_s = I_t^s\}$, where

$$I_t^s = \inf_{s \leq r \leq t} Y_r .$$

Under our assumptions however, this Lebesgue measure is always zero (if $s < t$, we have $P(Y_s = I_t^s) = 0$ because 0 is regular for $(-\infty, 0)$) and so we need to use some kind of local time that will measure the size of the set in consideration. More precisely, for a fixed $t > 0$, we introduce the time-reversed process

$$\hat{Y}_r^{(t)} = Y_t - Y_{(t-r)-} , \quad 0 \leq r \leq t \qquad (Y_{0-} = 0 \text{ by convention})$$

and its supremum process

$$\hat{S}_r^{(t)} = \sup_{0 \leq s \leq r} \hat{Y}_s^{(t)} , \quad 0 \leq r \leq t .$$

Note that $(\hat{Y}_r^{(t)}, \hat{S}_r^{(t)}; 0 \leq r \leq t)$ has the same distribution as $(Y_r, S_r; 0 \leq r \leq t)$. Via time-reversal, the set $\{s \leq t, Y_s = I_t^s\}$ corresponds to the set $\{s \leq t, \hat{S}_s^{(t)} = \hat{Y}_s^{(t)}\}$. This leads us to the following definition.

Definition. *For every $t \geq 0$, we let H_t be the local time at 0, at time t, of the process $\hat{S}^{(t)} - \hat{Y}^{(t)}$. The process $(H_t, t \geq 0)$ is called the height process.*

Obviously, the normalization of local time is the one that was described in Section 2. From the previous definition it is not clear that the sample paths of $(H_t, t \geq 0)$ have any regularity property. In order to avoid technical difficulties, we will reinforce assumption (1) by imposing that

$$\beta > 0 . \tag{1}'$$

We emphasize that (1)' is only for technical convenience and that all theorems and propositions that follow hold under (1) (for a suitable choice of a modification of $(H_t, t \geq 0)$).

Under (1)' we can get a simpler expression for H_t. Indeed from (4) we have

$$H_t = \frac{1}{\beta} m(\{\hat{S}_r^{(t)}, 0 \leq r \leq t\}) ,$$

or equivalently,

$$H_t = \frac{1}{\beta} m(\{I_t^r, 0 \leq r \leq t\}) . \tag{5}$$

The right side of the previous formula obviously gives a continuous modification of H (recall that Y has no negative jumps). From now on we deal only with this modification.

If $\psi(u) = \beta u^2$, Y is a (scaled) linear Brownian motion and has continuous paths. The previous formula then implies that $H_t = \frac{1}{\beta}(Y_t - I_t)$ is a (scaled) reflected Brownian motion, by a famous theorem of Lévy.

We can now state our main results. The key underlying idea is that H codes the genealogy of a ψ-CSBP in the same way as reflected Brownian motion codes the genealogy of the Feller diffusion. Our first theorem shows that the local time process of H (evaluated at a suitable stopping time), as a function of the space variable, is a ψ-CSBP.

Theorem 2. *For every $r > 0$, set $\tau_r = \inf\{t, I_t = -r\}$. There exists a ψ-CSBP $X = (X_a, a \geq 0)$ started at r, such that for every $h \in \mathcal{B}_{b+}(\mathbb{R}_+)$,*

$$\int_0^\infty da\, h(a) X_a = \int_0^{\tau_r} ds\, h(H_s) .$$

Obviously X can be defined by

$$X_a = \lim_{\varepsilon \downarrow 0} \frac{1}{\varepsilon} \int_0^{\tau_r} ds\, 1_{\{a < H_s < a+\varepsilon\}}, \text{ a.s.}$$

Remark. It is easy to verify that a.s. for every $t \geq 0$, $H_t = 0$ iff $Y_t = I_t$. (The implication $Y_t = I_t \implies H_t = 0$ is trivial.) Since τ_r is the inverse local time at 0 of $Y - I$, we can also interpret τ_r as the inverse local time at 0 of H. Indeed, Theorem 2 implies

$$r = X_0 = \lim_{\varepsilon \downarrow 0} \frac{1}{\varepsilon} \int_0^{\tau_r} ds\, 1_{\{0 < H_s < \varepsilon\}} , \text{ a.s.}$$

from which it easily follows that for every $t \geq 0$

$$\lim_{\varepsilon \downarrow 0} \frac{1}{\varepsilon} \int_0^t ds\, 1_{\{0 < H_s < \varepsilon\}} = -I_t , \text{ a.s.}$$

Using this remark, we see that the case $\psi(u) = \beta u^2$ of the previous theorem reduces to a classical Ray-Knight theorem on the Markovian properties of Brownian local times.

Our second theorem gives a snake-like construction of (ξ, ψ)-superprocesses. As in Chapter IV we consider a Markov process ξ with values in a Polish space E, satisfying the assumptions in Section IV.1. We also fix a point $x \in E$. We use the notation introduced in Section IV.1.

We then construct a process $(W_s, s \geq 0)$ with values in $\mathcal{W}_x$, whose law is characterized by the following two properties:

(i) $\zeta_s = \zeta_{W_s}$, $s \geq 0$ is distributed as the process H_s, $s \geq 0$.

(ii) Conditionally on $\zeta_s = f(s)$, $s \geq 0$, the process W has distribution Θ_x^f.

Note that this is exactly similar to the construction of Chapter IV, but the role of reflected Brownian motion is played by the processs H. There is another significant difference. The process W is not Markovian, because H itself is not. This explains why we constructed W started at the trivial path x and not with a general starting point $w \in \mathcal{W}_x$ (this would not make sense, see however the comments in the next section).

Arguments similar to the proof of Lemma IV.1 show that W is continuous in probability (see [LL2] for details). In particular we may and will choose a measurable modification of W.

Theorem 3. *There exists a (ξ, ψ)-superprocess $Z = (Z_a, a \geq 0)$, with $Z_0 = r\delta_x$, such that for every $h \in \mathcal{B}_{b+}(\mathbb{R}_+)$, $g \in \mathcal{B}_{b+}(E)$,*

$$\int_0^\infty h(a)\langle Z_a, g \rangle da = \int_0^{T_r} h(H_s)g(\hat{W}_s)ds .$$

This is clearly analogous to Theorem IV.4. Note however that we restricted our attention to a superprocess started at $r\delta_x$. As in Theorem IV.4, we could have obtained a general initial value for Z by introducing the excursion measures of W. These excursion measures are easily defined from the excursion measure of H away from 0, which is itself defined as the law of $(H_s, s \geq 0)$ under the excursion measure N of $Y - I$.

4   The exploration process

Before we proceed to the proofs, we need to introduce a crucial tool. We noticed that H is in general not a Markov process. For the calculations that will follow it is important to consider another process which contains more information than H and is Markovian.

Definition. *The exploration process $(\rho_t, t \geq 0)$ is the process with values in $\mathcal{M}_f(\mathbb{R}_+)$ defined by*

$$\langle \rho_t, g \rangle = \int_{[0,t]} d_s I_t^s \, g(H_s) ,$$

for $g \in \mathcal{B}_{b+}(\mathbb{R}_+)$. The integral in the right side is with respect to the increasing function $s \longrightarrow I_t^s$.

We can easily obtain a more explicit formula for ρ_t: A change of variables using (5) shows that

$$\langle \rho_t, g \rangle = \int_0^t ds\, I_t^s\, g\big(\beta^{-1}m(\{I_s^r, r \le s\})\big)$$

$$= \int_0^t ds\, I_t^s\, g\big(\beta^{-1}m(\{I_t^r, r \le s\})\big)$$

$$= \beta \int_0^{H_t} da\, g(a) + \sum_{s \le t:Y_{s-} < I_t^s} (I_t^s - Y_{s-})g(H_s)$$

so that

$$\rho_t(da) = \beta 1_{[0, H_t]}(a)da + \sum_{s \le t:Y_{s-} < I_t^s} (I_t^s - Y_{s-})\delta_{H_s}(da) . \qquad (6)$$

From this formula it is clear that

$$\operatorname{supp}\rho_t = [0, H_t] , \text{ for every } t \ge 0 , \text{ a.s.}$$

The definition of ρ_t also shows that

$$\langle \rho_t, 1 \rangle = Y_t - I_t .$$

The process $(\rho_t, t \ge 0)$ is the continuous analogue of the Markov chain $(\rho(n), n \ge 0)$ of Section 1.

If $\mu \in M_f(\mathbb{R}_+)$ and $a \in \mathbb{R}$ we define $k_a\mu \in M_f(\mathbb{R}_+)$ by the formula

$$k_a\mu([0, r]) = \mu([0, r]) \wedge a^+ .$$

When $a \le 0$, $k_a\mu = 0$, and when $a > 0$, $k_a\mu$ can be interpreted as the measure μ "truncated at mass a".

If $\mu \in M_f(\mathbb{R}_+)$ has compact support and $\nu \in M_f(\mathbb{R}_+)$, the concatenation $[\mu, \nu]$ is defined by

$$\int [\mu, \nu](dr)h(r) = \int \mu(dr)h(r) + \int \nu(dr)h\big(H(\mu) + r\big)$$

where $H(\mu) = \sup(\operatorname{supp}\mu)$.

Proposition 4. *The process $(\rho_t, t \ge 0)$ is a càdlàg strong Markov process with values in the space $M_f(\mathbb{R}_+)$ of all finite measures on $\mathbb{R}_+$. If $\theta \in M_f(\mathbb{R}_+)$, the process started at θ can be defined by the explicit formula*

$$\rho_t^\theta = [k_{<\theta,1>+I_t}\theta, \rho_t].$$

Proof. The càdlàg property of paths follows from the explicit formula (6). This formula shows more precisely that t is a discontinuity time for ρ iff it is so for Y, and $\rho_t = \rho_{t-} + \Delta Y_t \, \delta_{H_t}$.

Then, let T be a stopping time of the canonical filtration $(\mathcal{F}_t)_{t \geq 0}$ of Y. Consider the shifted process

$$Y_t^{(T)} = Y_{T+t} - Y_T, \quad t \geq 0,$$

which has the same distribution as Y and is independent of $\mathcal{F}_T$. Then, from the explicit formulas for ρ and H, one easily verifies that, a.s. for every $t > 0$,

$$\rho_{T+t} = [k_{<\rho_T,1>+I_t^{(T)}}\rho_T, \rho_t^{(T)}],$$

with an obvious notation for $\rho_t^{(T)}$ and $I_t^{(T)}$. The statement of Proposition 4 now follows from the fact that $(I_t^{(T)}, \rho_t^{(T)})$ has the same distribution as (I_t, ρ_t) and is independent of $\mathcal{F}_T$. □

Remark. In view of applications to superprocesses (cf Theorem 3 above), the right generalization of the Brownian snake of the previous chapters is the pair (ρ, W), which is Markovian in contrast to the process W alone. The process (ρ, W) is called the (ξ, ψ)-Lévy snake.

The following two propositions give properties of ρ that play a central role in the proof of Theorems 2 and 3. The first proposition gives an explicit formula for the invariant measure of ρ, and the second one describes the potential kernel of ρ killed when it hits 0.

Before stating these results, we give some important remarks. Recall that N denotes the excursion measure of $Y - I$ away from 0. Formulas (5) and (6) providing explicit expressions for the processes ρ and H still make sense under the excursion measure N. Furthermore, these formulas show that both ρ_t and H_t depend only on the values taken by $Y - I$ on the excursion e_t of $Y - I$ that straddles t, and

$$\rho_t = \rho_{t-a_t}(e_t), \qquad H_t = H_{t-a_t}(e_t),$$

if a_t denotes the starting time of this excursion. Since $\langle \rho_t, 1 \rangle = Y_t - I_t$ the excursion intervals of ρ away from 0 are the same as those of $Y - I$, and the "law" of $(\rho_t, t \geq 0)$ under N is easily identified with the excursion measure of the Markov process ρ away from 0.

We set $\psi^*(u) = \psi(u) - \alpha u$ and denote by $U = (U_t, t \geq 0)$ a subordinator with Laplace exponent ψ^*, i.e. with drift β and Lévy measure $\pi([r, \infty))dr$.

Proposition 5. *For every nonnegative measurable function Φ on $\mathcal{M}_f(\mathbb{R}_+)$,*

$$N\left(\int_0^\sigma dt \, \Phi(\rho_t) \right) = \int_0^\infty da \, e^{-\alpha a} \, E\big(\Phi(J_a)\big),$$

where $J_a(dr) = 1_{[0,a]}(r) \, dU_r$.

Proof. We may assume that Φ is bounded and continuous. From excursion theory for $Y - I$ and the remarks preceding the proposition, we have for every $\varepsilon > 0, C > 0$,

$$N\left(\int_0^\sigma dt\, \Phi(\rho_t)\, 1_{\{H_t \leq C\}}\right) = \frac{1}{\varepsilon} E\left(\int_0^{\tau_\varepsilon} dt\, \Phi(\rho_t)\, 1_{\{H_t \leq C\}}\right)$$

$$= \frac{1}{\varepsilon} \int_0^\infty dt\, E\left(1_{\{t < \tau_\varepsilon, H_t \leq C\}} \Phi(\rho_t)\right).$$

Then, for every fixed $t > 0$, we use time-reversal at time t. Recalling the definition of H and ρ, we see that

$$\rho_t = \hat{\eta}_t^{(t)}$$

where $\hat{\eta}_t^{(t)}$ is defined by

$$\langle \hat{\eta}_t^{(t)}, f \rangle = \int_0^t d\hat{S}_r^{(t)} f(\hat{L}_t^{(t)} - \hat{L}_r^{(t)})$$

and $\hat{L}_r^{(t)} = \beta^{-1} m(\{\hat{S}_s^{(t)}, 0 \leq s \leq r\})$ as in (4). Similarly,

$$\{t < \tau_\varepsilon, H_t \leq C\} = \{\hat{S}_t^{(t)} - \hat{Y}_t^{(t)} < \varepsilon, \hat{L}_t^{(t)} \leq C\}$$

and so we can write

$$E\left(1_{\{t < \tau_\varepsilon, H_t \leq C\}} \Phi(\rho_t)\right) = E\left(1_{\{\hat{S}_t^{(t)} - \hat{Y}_t^{(t)} < \varepsilon, \hat{L}_t^{(t)} \leq C\}} \Phi(\hat{\eta}_t^{(t)})\right)$$

$$= E\left(1_{\{S_t - Y_t < \varepsilon, L_t \leq C\}} \Phi(\eta_t)\right)$$

where

$$\langle \eta_t, f \rangle = \int_0^t dS_r\, f(L_t - L_r).$$

Summarizing, we have for every $\varepsilon > 0$

$$N\left(\int_0^\sigma dt\, \Phi(\rho_t)\, 1_{\{H_t \leq C\}}\right) = E\left(\frac{1}{\varepsilon} \int_0^\infty dt\, 1_{\{S_t - Y_t < \varepsilon, L_t \leq C\}} \Phi(\eta_t)\right).$$

Note from (2) that the random measures $\varepsilon^{-1} 1_{\{S_t - Y_t < \varepsilon\}} dt$ converge in probability to the measure dL_t. Furthermore, (2)′ allows us to pass to the limit under the expectation sign and we arrive at

$$\lim_{\varepsilon \to 0} E\left(\frac{1}{\varepsilon} \int_0^\infty dt\, 1_{\{S_t - Y_t < \varepsilon, L_t \leq C\}} \Phi(\eta_t)\right) = E\left(\int_0^\infty dL_t\, 1_{\{L_t \leq C\}} \Phi(\eta_t)\right)$$

$$= E\left(\int_0^{L_\infty \wedge C} da\, \Phi(\eta_{L^{-1}(a)})\right).$$

We finally let C tend to ∞ to get

$$N\left(\int_0^\sigma dt\, \Phi(\rho_t) \right) = E\left(\int_0^{L_\infty} da\, \Phi(\eta_{L^{-1}(a)}) \right).$$

Then note that, on the event $\{a < L_\infty\}$,

$$\langle \eta_{L^{-1}(a)}, f \rangle = \int_0^{L^{-1}(a)} dS_r\, f(a - L_r) = \int_0^a dV_s\, f(a - s),$$

where $V_s = S_{L^{-1}(s)}$ is a subordinator with exponent $\frac{\psi(\lambda)}{\lambda}$ (cf (3)). Hence, $P[a < L_\infty] = P[L^{-1}(a) < \infty] = e^{-\alpha a}$ and conditionally on $\{L^{-1}(a) < \infty\}$, $\eta_{L^{-1}(a)}$ has the same distribution as J_a, which completes the proof.   □

We denote by M the measure on $\mathcal{M}_f(\mathbb{R}_+)$ defined by:

$$\langle M, \Phi \rangle = \int_0^\infty da\, e^{-\alpha a}\, E\big(\Phi(J_a)\big).$$

It follows from Proposition 5 that the measure M is invariant for ρ (we will not need this fact in what follows).

Proposition 6. *Let $\theta \in \mathcal{M}_f(\mathbb{R}_+)$ and let ρ^θ be as in Proposition 4. Define $T_0^{(\theta)} = \inf\{s \geq 0, \rho_s^\theta = 0\}$. Then,*

$$E\left(\int_0^{T_0^{(\theta)}} ds\, \Phi(\rho_s^\theta) \right) = \int_0^{<\theta,1>} dr \int M(d\mu)\, \Phi([k_r\theta, \mu]).$$

Proof. First note that $T_0^{(\theta)} = \tau_{<\theta,1>}$ by an immediate application of the definition of ρ^θ (notice that $\rho_{\tau_a} = 0$ for every $a \geq 0$ a.s.). Then, denote by (a_j, b_j), $j \in J$ the excursion intervals of $Y - I$ away from 0 before time $\tau_{<\theta,1>}$, and by e_j, $j \in J$ the corresponding excursions. Note that $\{t \geq 0, Y_t = I_t\}$ has zero Lebesgue measure a.s., since, for every $t > 0$, $P(Y_t = I_t) = 0$ by time-reversal and the regularity of 0 for $(0, \infty)$. As we observed before Proposition 5, we have $\rho_s = \rho_{s-a_j}(e_j)$ for every $s \in (a_j, b_j)$, $j \in J$, a.s. It follows that

$$E\left(\int_0^{T_0^{(\theta)}} ds\, \Phi(\rho_s^\theta) \right) = E\left(\sum_{j \in J} \int_0^{b_j - a_j} dr\, \Phi([k_{<\theta,1>+I_{a_j}}\theta, \rho_r(e_j)]) \right).$$

By excursion theory, the point measure

$$\sum_{j \in J} \delta_{I_{a_j}, e_j}(dude)$$

is a Poisson point measure with intensity $1_{[-<\theta,1>,0]}(u)du\, N(de)$. Hence,

$$E\left(\int_0^{T_0^{(\theta)}} ds\, \Phi(\rho_s^\theta) \right) = \int_0^{<\theta,1>} du\, N\left(\int_0^\sigma dr\, \Phi([k_u\theta, \rho_r]) \right),$$

and the desired result follows from Proposition 5.   □

5   Proof of Theorem 2

We will now prove Theorem 2. Theorem 3 can be proved along the same lines (see [LL2]).

Let $h \in \mathcal{B}_{b+}(\mathbb{R}_+)$ with compact support. It is enough to prove that

$$E\left(\exp - \int_0^{T_r} ds\, h(H_s)\right) = E\left(\exp - \int_0^\infty da\, h(a)\, X_a\right), \qquad (7)$$

where X is a ψ-CSBP with $X_0 = r$. By Corollary II.9 (specialized to the case $f = 1$), the right side of (7) is equal to $\exp(-r\, w(0))$, where the function w is the unique nonnegative solution of the integral equation

$$w(t) + \int_t^\infty dr\, \psi(w(r)) = \int_t^\infty dr\, h(r). \qquad (8)$$

On the other hand, by excursion theory for the process $Y - I$ and the remarks preceding Proposition 5, we have

$$E\left[\exp - \int_0^{T_r} ds\, h(H_s)\right] = \exp -r\, N\left(1 - \exp - \int_0^\sigma ds\, h(H_s)\right).$$

Thus it suffices to verify that the function

$$w(t) = N\left(1 - \exp - \int_0^\sigma ds\, h(t + H_s)\right)$$

solves (8).

To this end, we will proceed in a way similar to the proof of Proposition IV.3 and expand the exponential in the definition of w. This leads to the calculation of the moments

$$T^n h(t) = \frac{1}{n!} N\left(\left(\int_0^\sigma ds\, h(t + H_s)\right)^n\right), \qquad n \geq 1.$$

Depending on the behavior of ψ near ∞, these moments may be infinite. Later we will make an assumption on ψ that guarantees the finiteness of the moments and the validity of the previously mentioned expansion. A suitable truncation can then be used to handle the general case.

To begin with, we observe that

$$T^n h(t) = N\left(\int_{\{0 < t_1 < \ldots < t_n < \sigma\}} dt_1 \ldots dt_n \prod_{i=1}^n h(t + H_{t_i})\right).$$

Note that $H_{t_i} = H(\rho_{t_i})$, where $H(\mu) = \sup(\operatorname{supp} \mu)$ as previously. The right-hand side of the preceding formula can be evaluated thanks to the following lemma. To simplify notation, we write $|\mu| = \langle \mu, 1 \rangle$ for $\mu \in M_f(\mathbb{R}_+)$.

Lemma 7. *For any nonnegative measurable functional F on $M_f(\mathbb{R}_+)^n$,*

$$
N\left(\int_{\{0 < t_1 < \ldots < t_n < \sigma\}} dt_1 \ldots dt_n \, F(\rho_{t_1}, \ldots, \rho_{t_n}) \right)
$$
$$
= \int Q^{(n)}(d\mu_1 \ldots d\mu_n da_2 \ldots da_n) \, F(\mu_1, \ldots, \mu_n),
$$

where $Q^{(n)}$ is the measure on $M_f(\mathbb{R}_+)^n \times \mathbb{R}_+^{n-1}$ which is defined by induction as follows. First, $Q^{(1)} = M$, and then $Q^{(n+1)}$ is the image of

$$
Q^{(n)}(d\mu_1 \ldots d\mu_n da_2 \ldots da_n) 1_{[0, |\mu_n|]}(a) da \, M(d\theta)
$$

under the mapping

$$
(\mu_1, \ldots, \mu_n, a_2, \ldots, a_n, a, \theta) \longrightarrow (\mu_1, \ldots, \mu_n, [k_a \mu_n, \theta], a_2, \ldots, a_n, a).
$$

Remark. This lemma is a generalization of Proposition III.3: In the case $\psi(u) = \beta u^2$, ρ_t is equal to β^{-1} times the Lebesgue measure on $[0, H_t]$, and the law of $(H_t, t \geq 0)$ under N is the Itô measure of Brownian excursions, up to a trivial scaling transformation.

Proof. The case $n = 1$ is Proposition 5. The proof is then completed by induction on n using the Markov property of ρ under N and Proposition 6. $\qquad \square$

By construction, the measures $\mu_1, \ldots, \mu_n$ exhibit a branching structure under $Q^{(n)}$ and the quantities $a_2, \ldots, a_n$ determine the levels of branching. We will examine this branching structure in detail and get a recursive relation between the measures $Q^{(n)}$, which is the key to the proof of Theorem 2.

For every $n \geq 1$, we set $\Theta^{(n)} = M_f(\mathbb{R}_+)^n \times \mathbb{R}_+^{n-1}$ $(\Theta^{(1)} = M_f(\mathbb{R}_+))$ and we take $\Theta = \cup_{n=1}^{\infty} \Theta^{(n)}$.

Let $n \geq 2$ and let $(\mu_1, \ldots, \mu_n, a_2, \ldots, a_n) \in \Theta^{(n)}$ be such that $a_j \leq |\mu_{j-1}| \wedge |\mu_j|$ and $k_{a_j} \mu_{j-1} = k_{a_j} \mu_j$ for every $j \in \{2, \ldots, n\}$ (these properties hold $Q^{(n)}$ a.e.). We define several quantities depending on $(\mu_1, \ldots, \mu_n, a_2, \ldots, a_n)$. First, we set

$$
b = \inf_{2 \leq j \leq n} a_j, \qquad h = H(k_b \mu_1).
$$

Notice that $k_b \mu_j = k_b \mu_1$ for every $j \in \{2, \ldots, n\}$.

We then set $b_- = \mu_1([0, h))$, $b_+ = \mu_1([0, h])$ and observe that $b_- \leq b \leq b_+$. We let $j_1 < j_2 < \cdots < j_{k-1}$ be the successive integers in $\{2, \ldots, n\}$ such that

$$a_{j_1} \in [b_-, b_+], \quad a_{j_2} \in [b_-, a_{j_1}], \quad \ldots \quad , a_{j_{k-1}} \in [b_-, a_{j_{k-2}}].$$

Here k is a (random) integer such that $2 \leq k \leq n$ and $b = a_{j_{k-1}}$ by construction. We also take $j_0 = 1$, $j_k = n + 1$ by convention. Informally, the integer k corresponds to the offspring number at the first branching, and h is the level of this branching.

We let ν_0 be the restriction of μ_1 (or of any μ_j) to $[0, h)$, and for every $j \in \{1, \ldots, n\}$, we define $\nu_j \in \mathcal{M}_f(\mathbb{R}_+)$ by taking $\nu_j([0, r]) = \mu_j((h, h + r])$. Fig. 1 illustrates the definition of the measures ν_j. In this figure, measures are represented by vertical segments, and the length of a segment corresponds to the total mass of the measure. The horizontal lines give the successive truncation levels (horizontal distances have no significance).
Finally, for every $l \in \{1, \ldots k\}$, we define

$$\Delta_l = (\mu_1^{(l)}, \ldots, \mu_{j_l - j_{l-1}}^{(l)}, a_2^{(l)}, \ldots, a_{j_l - j_{l-1}}^{(l)}) \in \Theta^{(j_l - j_{l-1})}$$

by setting

$$\mu_i^{(l)} = \nu_{j_{l-1} + i - 1}, \qquad\qquad 1 \leq i \leq j_l - j_{l-1},$$
$$a_i^{(l)} = a_{j_{l-1} + i - 1} - a_{j_{l-1}}, \quad 2 \leq i \leq j_l - j_{l-1},$$

where by convention $a_1 = b_+$.
The next lemma can be viewed as a generalization of Theorem III.4.

Lemma 8. *For every integer* $p \in \{2, \ldots, n\}$, *for every measurable subsets* $A_0, \ldots, A_n$ *of* Θ, *we have*

$$Q^{(n)}(k = p, \nu_0 \in A_0, \Delta_1 \in A_1, \ldots, \Delta_p \in A_p)$$
$$= \gamma_p M(A_0) \sum_{n_1 + \ldots + n_p = n, n_i \geq 1} Q^{(n_1)}(A_1) \ldots Q^{(n_p)}(A_p),$$

where $\gamma_p = \beta 1_{\{p=2\}} + \int \frac{y^p}{p!} \pi(dy)$.

Proof. We will derive Lemma 8 from a slightly more precise result, which is proved by induction on n. We keep the previous notation and also set $\Delta b = b_+ - b_-$, $c_i = a_{j_i} - b_-$ for $1 \leq i \leq k - 1$. Then, if B is a Borel subset of $\mathbb{R}_+^p$,

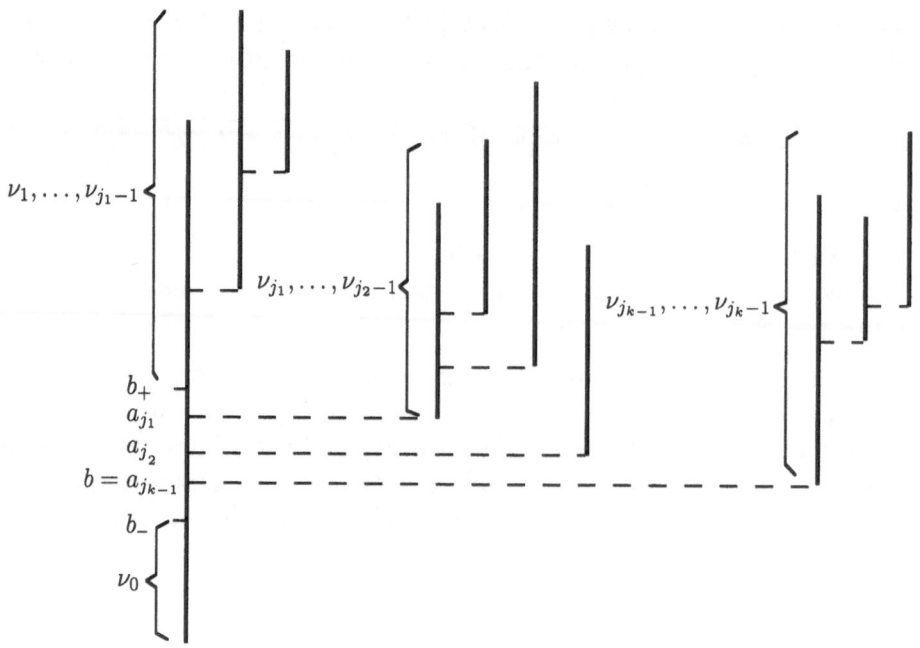

Fig. 1

we claim that

$$Q^{(n)}(k = p, (\Delta b, c_1, \ldots, c_{p-1}) \in B, \nu_0 \in A_0, \Delta_1 \in A_1, \ldots, \Delta_p \in A_p)$$
$$= \left(\beta 1_{\{p=2\}} 1_B(0,0) + \int \tilde{\pi}(dy) \int_0^y dz_1 \ldots \int_0^{z_{p-2}} dz_{p-1} 1_B(y, z_1, \ldots, z_{p-1}) \right)$$
$$\times \; M(A_0) \sum_{\substack{n_1+\ldots+n_p=n \\ n_i \geq 1}} Q^{(n_1)}(A_1) \ldots Q^{(n_p)}(A_p),$$

$$(9)$$

where $\tilde{\pi}(dy) = \pi([y, \infty)) \, dy$. Clearly, Lemma 8 follows from (9). Before proceeding to the proof of (9), we state a lemma giving the "law" under $M(d\mu)$ of the splitting of μ at a uniformly distributed mass level.

Lemma 9. *If $\mu \in M_f(\mathbb{R}_+)$ and $a \in (0, |\mu|)$, define $r = r(\mu, a)$ by $r = H(k_a \mu)$, and then $\tau_r \mu, \sigma_r \mu \in M_f(\mathbb{R}_+)$ by $\tau_r \mu = \mu_{|[0,r)}$, $\sigma_r \mu([0, u]) = \mu((r, r + u])$ for every $u \geq 0$. Then,*

$$\int M(d\mu) \int_0^{|\mu|} da \, F(\tau_r \mu, \sigma_r \mu, \mu(\{r\}), a - |\tau_r \mu|)$$
$$= \iint M(d\mu_1) M(d\mu_2) \left(\beta \, F(\mu_1, \mu_2, 0, 0) + \int \tilde{\pi}(dy) \int_0^y dz \, F(\mu_1, \mu_2, y, z) \right).$$

Proof. As in Proposition 5, write $U = (U_t, t \geq 0)$ for a subordinator with drift β and Lévy measure $\tilde{\pi}$. For every $a \geq 0$, set $\eta_a = \inf\{t, U_t \geq a\}$. By the definition of M, the left-hand side of the formula of Lemma 9 can be written as

$$\int_0^\infty dt\, e^{-\alpha t} \int_0^\infty da\, E\left[1_{\{a<U_t\}}\, F(1_{[0,\eta_a)}(s)dU_s, \sigma_{\eta_a}(1_{[0,t]}(s)dU_s), \Delta U_{\eta_a}, a - U_{\eta_a-})\right].$$

We may assume that F is of the form $F(\mu_1, \mu_2, u, v) = 1_{A_1}(\mu_1)1_{A_2}(\mu_2)1_B(u,v)$. Then the strong Markov property of U at time η_a shows that the previous expression is also equal to

$$E\left[\int_0^\infty da\, e^{-\alpha \eta_a}\, 1_{A_1}(1_{[0,\eta_a)}(s)dU_s)1_B(\Delta U_{\eta_a}, a - U_{\eta_a-})\right] M(A_2)$$

$$= M(A_2) \times \left(E\left[\sum_{t:\Delta U_t>0} e^{-\alpha t}1_{A_1}(1_{[0,t)}(s)dU_s)\int_0^{\Delta U_t} 1_B(\Delta U_t, z)\, dz\right]\right.$$

$$\left. + E\left[\int_0^\infty da\, 1_{\{\Delta U_{\eta_a}=0\}}e^{-\alpha \eta_a}1_{A_1}(1_{[0,\eta_a)}(s)dU_s)\, 1_B(0,0)\right]\right).$$

The first term of the sum inside parentheses is equal to

$$E\left[\int_0^\infty dt\, e^{-\alpha t}\, 1_{A_1}(1_{[0,t)}(s)dU_s)\int \tilde{\pi}(dy)\int_0^y 1_B(y,z)dz\right]$$

$$= \left(\int \tilde{\pi}(dy)\int_0^y 1_B(y,z)dz\right) M(A_1),$$

whereas the change of variable $s = \eta_a$ gives for the second term

$$E\left[\int_0^\infty dU_t\, 1_{\{\Delta U_t=0\}}e^{-\alpha t}1_{A_1}(1_{[0,t)}(s)dU_s)\, 1_B(0,0)\right] = \beta\, 1_B(0,0)\, M(A_1).$$

Lemma 9 follows. □

Proof of (9). First consider the case $n = 2$. Then necessarily $k = 2$. Furthermore, from the construction of $Q^{(2)}$, we have with the notation of Lemma 9,

$$Q^{(2)}\left(k = 2, (\Delta b, c_1) \in B, \nu_0 \in A_0, \nu_1 \in A_1, \nu_2 \in A_2\right)$$

$$= \iint M(d\mu)M(d\mu')\int_0^{|\mu|} da\, 1_B(\mu(\{r\}), a - |\tau_r\mu|)\, 1_{A_0}(\tau_r\mu)1_{A_1}(\sigma_r\mu)1_{A_2}(\mu')$$

$$= M(A_2)\int M(d\mu)\int_0^{|\mu|} da\, 1_B(\mu(\{r\}), a - |\tau_r\mu|)\, 1_{A_0}(\tau_r\mu)1_{A_1}(\sigma_r\mu)$$

$$= \left(\beta\, 1_B(0,0) + \int \tilde{\pi}(dy)\int_0^y dz\, 1_B(y,z)\right) M(A_0)M(A_1)M(A_2),$$

by Lemma 9. This gives the case $n = 2$.

To complete the proof, we argue by induction on n. Under $Q^{(n+1)}$ we have $\mu_{n+1} = [k_{a_{n+1}}\mu_n, \theta]$ for some $a_{n+1} \in [0, |\mu_n|]$, $\theta \in M_f(\mathbb{R}_+)$. To avoid confusion, write $k^{(n)}$, $b^{(n)}$, $b_-^{(n)}$, etc. for the quantities defined at the order n. We need to treat separately the following cases:

- If $a_{n+1} > b^{(n)}$, then $k^{(n+1)} = k^{(n)}$, and $\Delta_i^{(n+1)} = \Delta_i^{(n)}$ for $i = 1, \ldots, k^{(n)} - 1$, whereas $\Delta_{k^{(n)}}^{(n+1)}$ is obtained by adding one "branch" to $\Delta_{k^{(n)}}^{(n)}$.

- If $b_-^{(n)} < a_{n+1} < b^{(n)}$, then $k^{(n+1)} = k^{(n)} + 1$, $\Delta_i^{(n+1)} = \Delta_i^{(n)}$ for $i = 1, \ldots, k^{(n)}$, and $\Delta_{k^{(n)}+1}^{(n+1)}$ consists of only one "branch".

- If $a_{n+1} < b_-^{(n)}$, then $k^{(n+1)} = 2$ and $\Delta_2^{(n+1)}$ consists of only one "branch".

Starting from formula (9) at order n and examining carefully each of these cases one arrives at the formula at order $n+1$. We leave details to the reader. $\qquad\square$

We come back to the proof of Theorem 2. By Proposition 5,

$$T^1 h(t) = \int M(d\nu)\, h(t + H(\nu)) = \int_0^\infty dr\, e^{-\alpha r}\, h(t + r). \qquad (10)$$

By using Lemma 7 and then Lemma 8, we get for $n \geq 2$:

$$T^n h(t) = \int Q^{(n)}(d\mu_1 \ldots d\mu_n da_2 \ldots da_n) \prod_{i=1}^{n} h(t + H(\mu_i))$$

$$= \int Q^{(n)}(d\mu_1 \ldots d\mu_n da_2 \ldots da_n) \prod_{l=1}^{k} \Big(\prod_{i=1}^{j_l - j_{l-1}} h(t + H(\nu_0) + H(\mu_i^{(l)})) \Big)$$

$$= \sum_{p=2}^{n} \gamma_p \int M(d\nu) \sum_{n_1+\ldots+n_p=n, n_i \geq 1} \prod_{l=1}^{p} Q^{(n_l)} \Big(\prod_{i=1}^{n_l} h(t + H(\nu) + H(\mu_i)) \Big)$$

$$= \sum_{p=2}^{n} \gamma_p \sum_{n_1+\ldots+n_p=n, n_i \geq 1} T^1 \Big(\prod_{l=1}^{p} T^{n_l} h \Big)(t).$$

We have thus obtained the recursive relation:

$$T^n h = \sum_{p=2}^{n} \gamma_p \sum_{n_1+\ldots+n_p=n, n_i \geq 1} T^1 \Big(\prod_{l=1}^{p} T^{n_l} h \Big) \qquad (11)$$

(compare with formula (3) in Chapter IV).

To complete the proof, we first assume that $\operatorname{supp} \pi \subset [0, A]$ for some $A < \infty$. This implies that the numbers γ_p are finite. Furthermore, ψ is analytic on $\mathbb{R}$, and

$$\psi(u) = \alpha u + \sum_{p=2}^{\infty} (-1)^p \gamma_p u^p.$$

Let $B > 0$ be such that $h(t) = 0$ if $t \geq B$. The recursive relation (11) implies the existence of a constant $C < \infty$ such that, for every $n \geq 1$,

$$T^n h(t) \leq C^n 1_{[0,B]}(t). \tag{12}$$

To prove this bound, introduce the nonnegative function v that solves the integral equation

$$v(t) = \delta 1_{[0,B]}(t) + \int_t^{\infty} \overline{\psi}(v(s))\, ds,$$

where $\overline{\psi}(u) = \alpha u + \sum_{p=2}^{\infty} \gamma_p u^p$ and δ is a positive constant. Note that the function v is well defined and bounded provided that δ is small enough. Choose $\varepsilon > 0$ so that $\varepsilon T^1 h \leq \delta$. An easy induction argument using (11) and the integral equation for v shows that $\varepsilon^n T^n h \leq v$ for every $n \geq 1$. The bound (12) follows.

By (12), we have for $0 < \lambda < C^{-1}$,

$$\sum_{n=1}^{\infty} \frac{\lambda^n}{n!} N\left(\left(\int_0^{\sigma} h(t + H_s)\, ds\right)^n\right) < \infty.$$

If $w_\lambda(t) = N(1 - \exp -\lambda \int_0^{\sigma} ds\, h(t + H_s))$, we obtain from Fubini's theorem that, for $0 < \lambda < C^{-1}$,

$$w_\lambda(t) = \sum_{n=1}^{\infty} (-1)^{n-1} \lambda^n T^n h(t). \tag{13}$$

Set $\psi^*(u) = \psi(u) - \alpha u = \sum_{p=2}^{\infty} (-1)^p \gamma_p u^p$. Again by Fubini's theorem, we have

$$\int_0^{\infty} dr\, e^{-\alpha r} \psi^*(w_\lambda(t+r))$$

$$= \int_0^{\infty} dr\, e^{-\alpha r} \sum_{p=2}^{\infty} (-1)^p \gamma_p \left(\sum_{n=1}^{\infty} (-1)^{n-1} \lambda^n T^n h(t+r)\right)^p$$

$$= \int_0^{\infty} dr\, e^{-\alpha r} \sum_{p=2}^{\infty} (-1)^p \gamma_p \sum_{n_1,\ldots,n_p \geq 1} (-1)^{\sum n_i - p} \lambda^{\sum n_i} T^{n_1} h(t+r) \ldots T^{n_p} h(t+r)$$

$$= \sum_{n=2}^{\infty} (-1)^n \lambda^n \sum_{p=2}^{n} \gamma_p \sum_{n_1+\ldots+n_p=n, n_i \geq 1} T^1(T^{n_1}h \ldots T^{n_p}h)(t)$$

$$= \sum_{n=2}^{\infty} (-1)^n \lambda^n T^n h(t),$$

using (11) in the last equality, and (10) in the previous one. Comparing with (13) gives

$$w_\lambda(t) + \int_0^\infty dr \, e^{-ar} \psi^*(w_\lambda(t+r)) = \lambda T^1 h(t) = \lambda \int_0^\infty dr \, e^{-ar} h(t+r).$$

It is then easy to verify that this equation is equivalent to

$$w_\lambda(t) + \int_0^\infty \psi(w_\lambda(t+r)) \, dr = \lambda \int_0^\infty h(t+r) \, dr.$$

The latter equation holds a priori for $0 < \lambda < C^{-1}$. However an argument of analytic continuation allows us to extend it to every $\lambda > 0$. In particular, it holds for $\lambda = 1$, which gives the desired equation (8).

We finally explain the truncation procedure needed to get rid of our assumption that π is supported on $(0, A]$ for some $A < \infty$. For every integer $k \geq 1$, we let $\pi^{(k)}$ denote the restriction of π to $(0, k]$, and we set

$$\psi^{(k)}(\lambda) = \left(\alpha + \int_{(k,\infty)} r\pi(dr) \right) \lambda + \beta\lambda^2 + \int \pi^{(k)}(dr) \, (e^{-r\lambda} - 1 + r\lambda).$$

Notice that $\psi^{(k)} \downarrow \psi$ as $k \uparrow \infty$. The Lévy process with exponent $\psi^{(k)}$ can be embedded in the Lévy process with exponent ψ via a suitable time-change. To explain this embedding (under the excursion measure), we introduce the stopping times $U_j^{(k)}$, $j \geq 0$ and $T_j^{(k)}$, $j \geq 1$ defined inductively as follows:

$$U_0^{(k)} = 0,$$
$$T_j^{(k)} = \inf\{s \geq U_{j-1}^{(k)}, \Delta Y_s > k\}, \qquad j \geq 1,$$
$$U_j^{(k)} = \inf\{s \geq T_j^{(k)}, Y_s = Y_{T_j^{(k)}-}\}, \qquad j \geq 1.$$

We then let $\Gamma^{(k)}$ be the random set

$$\Gamma^{(k)} = \bigcup_{j=0}^{\infty} [U_j^{(k)}, T_{j+1}^{(k)})$$

and define $\eta_s^{(k)} = \int_0^s 1_{\Gamma^{(k)}}(r)dr$, $\gamma_s^{(k)} = \inf\{r, \eta_r^{(k)} > s\}$. Then, it is easy to verify that the process $Y_s^{(k)} := Y_{\gamma_s^{(k)}}$, $s \geq 0$ is distributed under N according to the excursion measure of the Lévy process with Laplace exponent $\psi^{(k)}$. Informally, $Y^{(k)}$ is obtained from Y by removing the jumps of size greater than k. Furthermore, our construction immediately shows that $H_s^{(k)} := H_{\gamma_s^{(k)}}$ is the height process associated with $Y^{(k)}$.

Set $\sigma^{(k)} = \eta_\sigma^{(k)}$. By the first part of the proof, we know that the function

$$w^{(k)}(t) = N\left(1 - \exp - \int_0^{\sigma^{(k)}} ds\, h(t + H_s^{(k)})\right)$$

solves (8) with ψ replaced by $\psi^{(k)}$. On the other hand, it is immediate that

$$w^{(k)}(t) = N\left(1 - \exp - \int_0^\sigma ds 1_{\Gamma^{(k)}}(s)h(t + H_s)\right)$$
$$\uparrow N\left(1 - \exp - \int_0^\sigma ds\, h(t + H_s)\right) = w(t)$$

as $k \uparrow \infty$. By simple monotonicity arguments, we conclude that w solves (8).
□

Bibliographical Notes

Chapter I

The classical books by Harris [Ha2], Athreya and Ney [AN] and Jagers [Ja] contain much about Galton-Watson branching processes and their generalizations. The idea of considering branching processes whose state space is "continuous" appears in Jirina [Ji]. Continuous-state branching processes and their connections with rescaled Galton-Watson processes were studied in the late sixties by Lamperti [La1], [La2] and Silverstein [Si] in particular. The convergence of rescaled critical Galton-Watson processes with finite variance towards the Feller diffusion had already been discussed by Feller [Fe]. Watanabe [Wa] used semigroup methods to construct a general class of measure-valued branching processes (later called superprocesses by Dynkin) including the one considered here. Watanabe also established a first result of approximation of superprocesses by branching particle systems. Similar approximations results have been obtained since by a number of authors in different settings: See in particular [Da1], [EK] (Chapter 9), [D3] and more recently [DHV]. For references concerning the other results mentioned in Chapter I, see below the notes about the corresponding chapters.

Chapter II

Theorem 1 is a special case of a result of Silverstein [Si] describing the general form of the Laplace exponent of a continuous-state branching process. Our construction of superprocesses via an approximation by branching particle systems is in the spirit of Dynkin [D3], [D4]. Lemma 6 is borrowed from [D3] (Lemma 3.1). Historical superprocesses were constructed independently by Dawson and Perkins [DP1], Dynkin [D3], [D4] and in a special case Le Gall [L3]. Regularity properties of superprocesses have been studied by Fitzsimmons [Fi1], [Fi2] via martingale methods (see also Dynkin [D2]). Proposition 7 and its proof are directly inspired from [D3] Lemma 4.1. The Laplace functional for the "weighted occupation time" of superprocesses (Corollary 9) was used by Iscoe [Is] to study properties of superprocesses in the quadratic branching case. Dawson's Saint-Flour lecture notes [Da2] provide a good survey of the literature about measure-valued processes until 1992. Dynkin's book [D8] gives a general presentation of the theory of superprocesses.

Chapter III

This chapter follows closely [L5], with some simplifications from Serlet [Se2]. Excursion theory was developed by Itô [It]. See the books [Bl], [RW] or [RY] for a detailed presentation of the Itô excursion measure. Our formalism for trees is in the spirit of Neveu [Ne]. The construction of branching trees embedded in linear Brownian motion or random walks has been the subject of many investigations. Harris [Ha1] first observed that the contour process of the critical geometric Galton-Watson tree is a positive excursion of simple random walk (see also [Dw]). Rogers [Ro] and Le Gall [L1] gave applications of this fact to properties of linear Brownian motion and its local times. Brownian analogues of Harris' observation were provided by Neveu and Pitman [NP1], [NP2], Le Gall [L2] and Abraham [Ab]. The coding of trees by functions was formalized by Aldous [Al3] (see also [L3]). Aldous' CRT was introduced and studied in [Al1] (see also [Al2] and [Al3]). The connection between the CRT and the normalized Brownian excursion (Theorem 6) was first obtained in [Al3], with a different method involving an approximation by conditioned Galton-Watson trees.

Chapter IV

The Brownian snake construction of superprocesses was first developed in [L3] with a slightly different approach. Our presentation here is more in the spirit of [L4] or [L6]. Proposition 2 giving the moment functionals for the Brownian snake was one of the motivation for the results of [L5] presented in Chapter III. Via the connection between the Brownian snake and quadratic superprocesses, these moment formulas also appear as consequences of the calculations in Dynkin [D1] (Theorem 1.1). Dynkin and Kuznetsov [DK1] used the main result of [L5] (our Theorem III.4) to prove an isomorphism theorem between Brownian snakes and superprocesses which is more precise than Theorem IV.4. The fact that a superprocess started at a general initial value can be written as a Poisson sum whose intensity involves the so-called "excursion measures" was observed in [EKR] and used in particular in [DP1] (Chapter 3). Many remarkable sample path properties of super-Brownian motion (much more precise than Corollary 9) have been established by Perkins and others: See e.g. [Su], [Pe1], [Pe2], [Pe3], [DIP], [Tr1], [AL], [Se1], [Se2], [LP], [DL2], [L12], [De]. The result of the exercise at the end of Section 5 is due to Tribe [Tr2]. Integrated super-Brownian excursion was discussed by Aldous [Al4] as a tree-based model for random distribution of mass.

Chapter V

Exit measures of superprocesses were introduced and studied by Dynkin [D3], [D4] for superprocesses with a general branching mechanism. In particular, the basic Theorem 4 is a special case of Dynkin's results. Our presentation follows [L6], which gives the very useful Lemma 5. This lemma also plays an important role in the obtention of sample path properties of super-Brownian motion: See in particular [LP] and [L12]. The probabilistic solution of the nonlinear problem (again for more general branching mechanisms) was derived in Dynkin [D5] (see also [D6] for analogous results in the parabolic setting). Lemma 7 is borrowed from the appendix to [D5], and Corollary 8 is (a special case of) the "mean value property" observed by Dynkin. The analytic part of Proposition 9 is not a difficult result and had been known for a long time, but the probabilistic approach is especially simple.

Chapter VI

The probabilistic representations of solutions with boundary blow-up (Propositions 1 and 2) are due to Dynkin [D5] (see also [D6] and [D7] for many related results including the parabolic setting). The existence of such solutions had been known for a long time by analytic methods: See Keller [Ke] and Osserman [Os]. The question of characterizing polar sets was first addressed by Perkins [Pe3], who showed that a set is not polar if it has nonzero capacity. The first part of the proof of Theorem 4 is an adaptation of Perkins' argument (see [L4] for a different approach using the potential theory of symmetric Markov processes). The converse was obtained by Dynkin [D5], who generalized the result to the case of stable branching mechanism. Dynkin's approach consists in observing that polar sets exactly correspond to removable singularities for the associated partial differential equation, and then using the characterization due to Baras and Pierre [BP]. The key duality argument of the proof presented here is also taken from [BP]. See Adams and Hedberg [AH] for a thorough discussion of Sobolev capacities and equivalent definitions. Results analogous to Theorem 4 in the parabolic setting are presented in [D6] and [D7]. The paper [L7] gives explicit calculations of certain capacitary distributions for the Brownian snake, which yield in particular the law of the process at the first time when it hits a nonpolar compact subset K (this law is described as the distribution of the solution to a certain stochastic differential equation). Section 3 is adapted from [DL1] with some simplifications. The problem of finding sufficient conditions for the existence of solutions of $\Delta u = u^p$ with boundary blow-up has been tackled by several authors in the analytic literature: See in particular [BM], [V1] and [MV1]. Theorem 6 and Corollary 7 have been extended by Delmas and Dhersin [DD] to a parabolic setting. Theorem 9 is an improvement of a result in [L6], along the lines of [Dh], Chapter 3. Lemma 10 was proved in [L8] and applied there to certain estimates of hitting probabilities for super-Brownian motion. See [MV1] for an analytic approach to the uniqueness of the nonnegative solution of $\Delta u = u^p$ in a domain (until now, this analytic approach requires more stringent conditions than the one of Theorem 9). The recent book by Véron [V2] is a good source of analytic references for the problems treated in this chapter and the next one.

Chapter VII

Proposition 1 is taken from [L7] (Proposition 4.4). Theorem II.8.1 of [D7] is a closely related result valid for the more general equation $\Delta u = u^\alpha$, $1 < \alpha \leq 2$. Theorem 2 characterizing boundary polar sets was conjectured by Dynkin [D7] and proved in [L7] (for the "easy" part) and [L9]. Some partial analytic results in this direction had been obtained previously by Gmira and Véron [GV] and Sheu [Sh1]. Dynkin and Kuznetsov [DK2] have extended Theorem 2 to equation $\Delta u = u^\alpha$, $1 < \alpha \leq 2$. Sections 2 and 3 follow closely [L11], except for Lemma 7 which was proved in [AL]. In the special case of equation $\Delta u = u^2$, Theorem 5 explains a phenomenon of nonuniqueness observed by Kondratyev and Nikishkin [KN] for singular solutions of $\Delta u = u^\alpha$ in a domain. References to the more recent work on trace problems for $\Delta u = u^\alpha$ in a domain of $\mathbb{R}^d$ are given in Section VII.4.

Chapter VIII

This chapter is based on the papers [LL1], [LL2]. Bertoin's book [Be] (especially Chapter VII) contains the basic facts about Lévy processes that are used in this chapter, with the exception of (2) or (2)' that can be found in [DuL]. The discrete construction of Section 1 is adapted from [LL1], but several other papers present closely related results, and use the random walk representation to get information on the asymptotic behavior of Galton-Watson processes: See in particular Borovkov and Vatutin [BV] and Bennies and Kersting [BK]. Theorem 3 is proved in detail in [LL2], where the details of the proof of Lemma 8 can also be found. Another more ancient connection between the ψ-continuous-state branching process and the Lévy process with Laplace exponent ψ had been observed by Lamperti [La2]. The paper [BLL] presents a different approach (based on subordination) to a snake-like construction of superprocesses with a general branching mechanism. Still another approach to the genealogical structure of superprocesses, and more general measure-valued processes, has been developed by Donnelly and Kurtz [DK1], [DK2]. The monograph [DuL] will give various applications of the results of this chapter.

Bibliography

[Ab] ABRAHAM, R. (1992) Un arbre aléatoire infini associé à l'excursion brownienne. Séminaire de Probabilités XXVI. *Lecture Notes Math.* **1526**, pp. 374–397. Springer, Berlin.

[AL] ABRAHAM, R., LE GALL, J.F. (1994) Sur la mesure de sortie du super-mouvement brownien. *Probab. Th. Rel. Fields* **99**, 251–275.

[AP] ADAMS, D.R., HEDBERG, L.I. (1996) *Function spaces and potential theory.* Springer, Berlin.

[Al1] ALDOUS, D. (1991) The continuum random tree I. *Ann. Probab.* **19**, 1–28.

[Al2] ALDOUS, D. (1991) The continuum random tree II: An overview. In: *Stochastic Analysis* (M.T. Barlow, N.H. Bingham eds), pp. 23–70. Cambridge University Press, Cambridge.

[Al3] ALDOUS, D. (1993) The continuum random tree III. *Ann. Probab.* **21**, 248–289.

[Al4] ALDOUS, D. (1993) Tree-based models for random distribution of mass. *J. Stat. Phys.* **73**, 625–641.

[AN] ATHREYA, K.B., NEY, P.E. (1972) *Branching processes.* Springer, Berlin.

[BM] BANDLE, C., MARCUS, M. (1992) Large solutions of semilinear elliptic equations: Existence, uniqueness and asymptotic behavior. *J. Analyse Math.* **58**, 8–24.

[BP] BARAS, P., PIERRE, M. (1984) Singularités éliminables pour des équations semilinéaires. *Ann. Inst. Fourier* **34**, 185–206.

[BK] BENNIES, J., KERSTING, G. (1997) A random walk approach to Galton-Watson trees. Preprint.

[Be] BERTOIN, J. (1996) *Lévy processes.* Cambridge University Press, Cambridge.

[Bl] BLUMENTHAL, R.M. (1992) *Excursions of Markov processes.* Birkhäuser, Boston.

[BLL] BERTOIN, J., LE GALL, J.F., LE JAN, Y. (1997) Spatial branching processes and subordination. *Canadian J. Math.* **49**, 24–54.

[BV] BOROVKOV, K.A., VATUTIN, V.A. (1996) On distribution tails and expectations of maxima in critical branching processes. *J. Appl. Probab.* **33**, 614–622.

[BCL] BRAMSON, M., COX, J.T., LE GALL, J.F. (1999) Coalescing random walks, the voter model and super-Brownian motion. In preparation.

[BG] BRAMSON, M., GRIFFEATH, D. (1980) Asymptotics for interacting particle systems on $\mathbb{Z}^d$. *Z. Wahrsch. verw. Gebiete* **53**, 183–196.

[CDP] COX, J.T., DURRETT, R., PERKINS, E.A. (1999) Rescaled voter models converge to super-Brownian motion. Preprint.

[Da1] DAWSON, D.A. (1975) Stochastic evolution equations and related measure processes. *J. Multivariate Anal.* **3**, 1–52.

[Da2] DAWSON, D.A. (1993) Measure-valued Markov processes. Ecole d'Eté de Probabilités de Saint-Flour 1991. *Lecture Notes Math.* **1541**, pp. 1–260. Springer, Berlin.

[DHV] DAWSON, D.A., HOCHBERG, K.J., VINOGRADOV, V. (1996) On weak convergence of branching particle systems undergoing spatial motion. In: *Stochastic analysis: Random fields and measure-valued processes*, pp. 65–79. Israel Math. Conf. Proc. 10. Bar-Ilan Univ., Ramat Gan.

[DIP] DAWSON, D.A., ISCOE, I., PERKINS, E.A. (1989) Super-Brownian motion: Path properties and hitting probabilities. *Probab. Th. Rel. Fields* **83**, 135–205.

[DP1] DAWSON, D.A., PERKINS, E.A. (1991) Historical processes. *Memoirs Amer. Math. Soc.* **454**.

[DP2] DAWSON, D.A., PERKINS, E.A. (1996) Measure-valued processes and stochastic partial differential equations. *Can. J. Math.* (special anniversary volume), 19–60.

[DP3] DAWSON, D.A., PERKINS, E.A. (1997) Measure-valued processes and renormalization of branching particle systems. In: *Stochastic partial differential equations: Six perspectives* (eds R. Carmona, B. Rozovski) AMS Mathematical Survey and Monographs Vol. **64**, pp. 45–106. AMS, Providence.

[De] DELMAS, J.F. (1998) Some properties of the range of super-Brownian motion. *Probab. Th. Rel. Fields*, to appear.

[DD] DELMAS, J.F., DHERSIN, J.S. (1998) Characterization of G-regularity for super-Brownian motion and consequences for parabolic partial differential equations. *Ann. Probab.*, to appear.

[DS] DERBEZ, E., SLADE, G. (1998) The scaling limit of lattice trees in high dimensions. *Commun. Math. Phys.* **193**, 69–104.

[Dh] DHERSIN, J.S. (1997) *Super-mouvement brownien, serpent brownien et équations aux dérivées partielles.* Thèse Université Paris VI.

[DL1] DHERSIN, J.S., LE GALL, J.F. (1997) Wiener's test for super-Brownian motion and for the Brownian snake. *Probab. Th. Rel. Fields* **108**, 103–129.

[DL2] DHERSIN, J.S., LE GALL, J.F. (1998) Kolmogorov's test for super-Brownian motion. *Ann. Probab.* **26**, 1041–1056.

[DK1] DONNELLY, P., KURTZ, T.G. (1996) A countable representation of the Fleming-Viot measure-valued branching diffusion. *Ann. Probab.* **24**, 1–16.

[DK2] DONNELLY, P., KURTZ, T.G. (1999) Particle representations for measure-valued population models. *Ann. Probab.* **27**, 166–205.

[DuL] DUQUESNE, T., LE GALL, J.F. (1999) Branching trees, Lévy processes and Lévy snakes. In preparation.

[DuP] DURRETT, R., PERKINS, E.A. (1998) Rescaled contact processes converge to super-Brownian motion for $d \geq 2$. Preprint.

[Dw] DWASS, M. (1975) Branching processes in simple random walk. *Proc. Amer. Math. Soc.* **51**, 251–274.

[D1] DYNKIN, E.B. (1988) Representation for functionals of superprocesses by multiple stochastic integrals, with applications to self-intersection local times. *Astérisque* **157–158**, 147–171.

[D2] DYNKIN, E.B. (1989) Regular transition functions and regular superprocesses. *Trans. Amer. Math. Soc.* **316**, 623–634.

[D3] DYNKIN, E.B. (1991) Branching particle systems and superprocesses. *Ann. Probab.* **19**, 1157–1194.

[D4] DYNKIN, E.B. (1991) Path processes and historical superprocesses. *Probab. Th. Rel. Fields* **90**, 89–115.

[D5] DYNKIN, E.B. (1991) A probabilistic approach to one class of nonlinear differential equations. *Probab. Th. Rel. Fields* **89**, 89–115.

[D6] DYNKIN, E.B. (1992) Superdiffusions and parabolic nonlinear differential equations. *Ann. Probab.* **20**, 942–962.

[D7] DYNKIN, E.B. (1993) Superprocesses and partial differential equations. *Ann. Probab.* **21**, 1185–1262.

[D8] DYNKIN, E.B. (1994) *An introduction to branching measure-valued processes.* CRM Monograph Series Vol. 6. Amer. Math. Soc., Providence.

[DK1] DYNKIN, E.B., KUZNETSOV, S.E. (1995) Markov snakes and superprocesses. *Probab. Th. Rel. Fields* **103**, 433–473.

[DK2] DYNKIN, E.B., KUZNETSOV, S.E. (1996) Superdiffusions and removable singularities for quasilinear partial differential equations. *Comm. Pure Appl. Math.* **49**, 125–176.

[DK3] DYNKIN, E.B., KUZNETSOV, S.E. (1996) Solutions of $Lu = u^\alpha$ dominated by L-harmonic functions. *J. Analyse Math.* **68**, 15–37.

[DK4] DYNKIN, E.B., KUZNETSOV, S.E. (1998) Trace on the boundary for solutions of nonlinear differential equations. *Trans. Amer. Math. Soc.* **350**, 4499–4519.

[DK5] DYNKIN, E.B., KUZNETSOV, S.E. (1998) Solutions of nonlinear differential equations on a Riemannian manifold and their trace on the Martin boundary. *Trans. Amer. Math. Soc.* **350**, 4521–4552.

[DK6] DYNKIN, E.B., KUZNETSOV, S.E. (1998) Fine topology and fine trace on the boundary associated with a class of semilinear differential equations. *Comm. Pure Appl. Math.* **51**, 897–936.

[DK7] DYNKIN, E.B., KUZNETSOV, S.E. (1999) Extinction of superdifffusions and semilinear partial differential equations. *J. Funct. Anal.* **162**, 346–378.

[DK8] DYNKIN, E.B., KUZNETSOV, S.E. (1998) Solutions of a class of semilinear differential equations: Trace and singularities on the boundary. Preprint.

[EKR] EL KAROUI, N., ROELLY, S. (1991) Propriétés de martingales, explosion et représentation de Lévy-Khintchine d'une classe de processus de branchement à valeurs mesures. *Stoch. Process. Appl.* **38**, 239–266.

[Et] ETHERIDGE, A.M. (1996) A probabilistic approach to blow-up of a semilinear heat equation. *Proc. Roy. Soc. Edinburgh Sect. A* **126**, 1235–1245.

[EK] ETHIER, S.N., KURTZ, T.G. (1985) *Markov processes: Characterization and convergence.* Wiley, New York.

[Fe] FELLER, W. (1951) Diffusion processes in genetics. *Proc. Second Berkeley Symp. Math. Statist. Prob.*, University of California Press, Berkeley, pp. 227–246.

[Fi1] FITZSIMMONS, P.J. (1988) Construction and regularity of measure-valued branching processes. *Israel J. Math.* **64**, 337–361.

[Fi2] FITZSIMMONS, P.J. (1992) On the martingale problem for measure-valued Markov branching processes. In: *Seminar on Stochastic Processes 1991*, pp. 39–51. Birkhäuser, Boston.

[GV] GMIRA, A., VÉRON, L. (1991) Boundary singularities of some nonlinear elliptic equations. *Duke Math. J.* **64**, 271–324.

[Ha1] HARRIS, T.E. (1952) First passage and recurrence distributions. *Trans. Amer. Math. Soc.* **73**, 471–486.

[Ha2] HARRIS, T.E. (1963) *The theory of branching processes.* Springer, Berlin.

[HS] HARA, T., SLADE, G. (1998) The incipient infinite cluster in high-dimensional percolation. *Electronic Res. Announc. Amer. Math. Soc.* **4**, 48–55.

[Is] ISCOE, I. (1986) A weighted occupation time for a class of measure-valued branching processes. *Probab. Th. Rel. Fields* **71**, 85–116.

[IL] ISCOE, I., LEE, T.Y. (1993) Large deviations for occupation times of measure-valued branching Brownian motions. *Stoch. Stoch. Rep.* **45**, 177–209.

[It] ITÔ, K. (1970) Poisson point processes attached to Markov processes. *Proc. Sixth Berkeley Symp. Math. Stat. Prob.*, vol. 3. University of California, Berkeley, pp. 225–239.

[Ja] JAGERS, P. (1975) *Branching processes with biological applications.* Wiley.

[Ji] JIRINA, M. (1957) Stochastic branching processes with continuous state space. *Czechosl. Math. J.* **8**, 292–313.

[Ka] KALLENBERG, O. (1986) *Random measures.* Academic Press, New York.

[Ke] KELLER, J.B. (1957) On solutions of $\Delta u = f(u)$. *Comm. Pure Appl. Math.* **10**, 503–510.

[KN] KONDRATYEV, V.A., NIKISHKIN, V.A. (1993) On positive solutions of singular boundary value problems for the equation $\Delta u = u^k$. *Russian J. Math. Physics* **1**, 131–135.

[KS] KONNO, N., SHIGA, T. (1988) Stochastic differential equations for some measure-valued diffusions. *Probab. Th. Rel. Fields* **79**, 201–225.

[Ku] KUZNETSOV, S.E. (1998) σ-moderate solutions of $Lu = u^\alpha$ and fine trace on the boundary. *C.R. Acad. Sci. Paris, Série I* **326**, 1189–1194.

[La1] LAMPERTI, J. (1967) The limit of a sequence of branching processes. *Z. Wahrsch. verw. Gebiete* **7**, 271–288.

[La2] LAMPERTI, J. (1967) Continuous-state branching processes. *Bull. Amer. Math. Soc.* **73**, 382–386.

[Le] LEE, T.Y. (1993) Some limit theorems for super-Brownian motion and semilinear partial differential equations. *Ann. Probab.* **21**, 979–995.

[L1] LE GALL, J.F. (1986) Une approche élémentaire des théorèmes de décomposition de Williams. *Séminaire de Probabilités XX. Lecture Notes Math.* **1204**, pp. 447–464. Springer.

[L2] LE GALL, J.F. (1989) Marches aléatoires, mouvement brownien et processus de branchement. *Séminaire de Probabilités XXIII. Lecture Notes Math.* **1372**, pp. 258–274. Springer.

[L3] LE GALL, J.F. (1991) Brownian excursions, trees and measure-valued branching processes. *Ann. Probab.* **19**, 1399–1439.

[L4] LE GALL, J.F. (1993) A class of path-valued Markov processes and its applications to superprocesses. *Probab. Th. Rel. Fields* **95**, 25–46.

[L5] LE GALL, J.F. (1993) The uniform random tree in a Brownian excursion. *Probab. Th. Rel. Fields* **96**, 369–383.

[L6] LE GALL, J.F. (1994) A path-valued Markov process and its connections with partial differential equations. In: *Proc. First European Congress of Mathematics, Vol.II*, pp. 185–212. Birkhäuser, Boston.

[L7] LE GALL, J.F. (1994) Hitting probabilities and potential theory for the Brownian path-valued process. *Ann. Inst. Fourier* **44**, 277–306.

[L8] LE GALL, J.F. (1994) A lemma on super-Brownian motion with some applications. In: *The Dynkin Festschrift* (M.I. Freidlin ed.), pp. 237–251. Birkhäuser, Boston.

[L9] LE GALL, J.F. (1995) The Brownian snake and solutions of $\Delta u = u^2$ in a domain. *Probab. Th. Rel. Fields* **102**, 393–432.

[L10] LE GALL, J.F. (1996) A probabilistic approach to the trace at the boundary for solutions of a semilinear parabolic partial differential equation. *J. Appl. Math. Stoch. Anal.* **9**, 399–414.

[L11] LE GALL, J.F. (1997) A probabilistic Poisson representation for positive solutions of $\Delta u = u^2$ in a domain. *Comm. Pure Appl. Math.* **50**, 69–103.

[L12] LE GALL, J.F. (1998) The Hausdorff measure of the range of super-Brownian motion. To appear in a special volume in honor of Harry Kesten.

[LL1] LE GALL, J.F., LE JAN, Y. (1998) Branching processes in Lévy processes: The exploration process. *Ann. Probab.* **26**, 213–252.

[LL2] LE GALL, J.F., LE JAN, Y. (1998) Branching processes in Lévy processes: Laplace functionals of snakes and superprocesses. *Ann. Probab.* **26**, 1407–1432.

[LP] LE GALL, J.F., PERKINS, E.A. (1995) The Hausdorff measure of the support of two-dimensional super-Brownian motion. *Ann. Probab.* **23**, 1719–1747.

[LPT] LE GALL, J.F., PERKINS, E.A., TAYLOR, S.J. (1995) The packing measure of the support of super-Brownian motion. *Stoch. Process. Appl.* **59**, 1–20.

[MV1] MARCUS, M., VÉRON, L. (1997) Uniqueness and asymptotic behavior of solutions with boundary blow up for a class of nonlinear elliptic equations. *Ann. Inst. H. Poincaré Anal. Non Linéaire* **14**, 237–274.

[MV2] MARCUS, M., VÉRON, L. (1998) The boundary trace of positive solutions of semilinear equations I: The subcritical case. *Arch. Rational Mech. Anal.* **144**, 201–231.

[MV3] MARCUS, M., VÉRON, L. (1998) The boundary trace of positive solutions of semilinear equations II: The supercritical case. *J. Math. Pures Appl.* **77**, 481–524.

[MP] MUELLER, C., PERKINS, E.A. (1992) The compact support property for solutions of the heat equation with noise. *Probab. Th. Rel. Fields* **93**, 325–358.

[Ne] NEVEU, J. (1986) Arbres et processus de Galton-Watson. *Ann. Inst. Henri Poincaré* **22**, 199–207.

[NP1] NEVEU, J., PITMAN, J.W. (1989) Renewal property of the extrema and tree property of the excursion of a one-dimensional Brownian motion. Séminaire de Probabilités XXIII. *Lecture Notes Math.* **1372**, pp. 239–247. Springer.

[NP2] NEVEU, J., PITMAN, J.W. (1989) The branching process in a Brownian excursion. Séminaire de Probabilités XXIII. *Lecture Notes Math.* **1372**, pp. 248–257. Springer.

[Os] OSSERMAN, R. (1957) On the inequality $\Delta u \geq f(u)$. *Pacific J. Math.* **7**, 1641–1647.

[Pa] PARTHASARATHY, K.R. (1967) *Probability measures on metric spaces.* Academic Press, New York.

[Pe1] PERKINS, E.A. (1988) A space-time property of a class of measure-valued branching diffusions. *Trans. Amer. Math. Soc.* **305**, 743–795.

[Pe2] PERKINS, E.A. (1989) The Hausdorff measure of the closed support of super-Brownian motion. *Ann. Inst. H. Poincaré Probab. Stat.* **25**, 205–224.

[Pe3] PERKINS, E.A. (1990) Polar sets and multiple points for super-Brownian motion. *Ann. Probab.* **18**, 453–491.

[PS] PORT, S.C., STONE, C.J. (1978) *Brownian motion and classical potential theory.* Academic, New York.

[RY] REVUZ, D., YOR, M. (1991) *Continuous martingales and Brownian motion.* Springer, Berlin.

[Re] REIMERS, M. (1989) One-dimensional stochastic partial differential equations and the branching measure diffusion. *Probab. Th. Rel. Fields* **81**, 319–340.

[Ro] ROGERS, L.C.G. (1984) Brownian local times and branching processes. Séminaire de Probabilités XVIII. *Lecture Notes Math.* **1059**, pp. 42–55. Springer.

[RW] ROGERS, L.C.G., WILLIAMS, D. (1987) *Diffusions, Markov processes and martingales, Vol.2.* Wiley, Chichester.

[Se1] SERLET, L. (1995) Some dimension results for super-Brownian motion. *Probab. Th. Rel. Fields* **101**, 371–391.

[Se2] SERLET, L. (1995) On the Hausdorff measure of multiple points and collision points of super-Brownian motion. *Stochastics Stoch. Rep.* **54**, 169–198.

[Sh1] SHEU, Y.C. (1994) Removable boundary singularities for solutions of some nonlinear differential equations. *Duke Math. J.* **74**, 701–711.

[Sh2] SHEU, Y.C. (1995) On positive solutions of some nonlinear differential equations: A probabilistic approach. *Stoch. Process. Appl.* **59**, 43–54.

[Sh3] SHEU, Y.C. (1997) Lifetime and compactness of range for super-Brownian motion with a general branching mechanism. *Stoch. Process. Appl.* **70**, 129–141.

[Si] SILVERSTEIN, M.L. (1968) A new approach to local times. *J. Math. Mech.* **17**, 1023–1054.

[Su] SUGITANI, S. (1987) Some properties for the measure-valued diffusion process. *J. Math. Soc. Japan* **41**, 437–462.

[Tr1] TRIBE, R. (1991) The connected components of the closed support of super-Brownian motion. *Probab. Th. Rel. Fields* **89**, 75–87.

[Tr2] TRIBE, R. (1992) The behavior of superprocesses near extinction. *Ann. Probab.* **20**, 286–311.

[V1] VÉRON, L. (1992) Semilinear elliptic equations with uniform blow-up at the boundary. *J. Analyse Math.* **59**, 231–250.

[V2] VÉRON, L. (1996) *Singularities of solutions of second order quasilinear equations.* Pitman Research Lecture Notes in Math. 353. Longman, Harlow.

[Wa] WATANABE, S. (1969) A limit theorem of branching processes and continuous state branching processes. *J. Math. Kyoto Univ.* **8**, 141–167.

Index